DARWINIAN POLITICS

THE RUTGERS SERIES
IN HUMAN EVOLUTION

Robert Trivers, *Founding Editor*

Lee Cronk, *Associate Editor*

Helen Fisher, *Advisory Editor*

Lionel Tiger, *Advisory Editor*

DARWINIAN POLITICS

The Evolutionary Origin of Freedom

Paul H. Rubin

RUTGERS UNIVERSITY PRESS
New Brunswick, New Jersey, and London

Library of Congress Cataloging-in-Publication Data

Rubin, Paul H.
    Darwinian politics : the evolutionary origin of freedom / Paul H. Rubin.
        p.   cm.
    Includes bibliographical references and index.
    ISBN 0-8135-3095-4 (cloth : alk. paper) — ISBN 0-8135-3096-2 (pbk. : alk. paper)
    1. Biopolitics.   2. Social Darwinism.   I. Title.

    JA80 .R93 2002
    306.2—dc21

                                                                    2001058677

British Cataloging-in-Publication information is available from the British Library.

Manufactured in the United States of America

To my exceptional wife, Martie Moss,
who has always had confidence in me

Contents

Preface

The notion that humans are born as blank slates (*tabula rasa,* to use Locke's Latin phrase) is no longer intellectually respectable among serious people. The big break came with the publication of E. O. Wilson's magisterial *Sociobiology* in 1975 and the founding of the discipline of the same name. This discipline has metamorphosed into what is now called evolutionary psychology, the branch of science based on the idea that our mental proclivities are rooted in our biological and evolutionary history. An important source for this notion is the 1992 book edited by Jerome H. Barkow, Leda Cosmides, and John Tooby, *The Adapted Mind.* Numerous other books have made this new learning more generally available, including such works as economist Robert Frank (1988), *Passions Within Reasons: The Strategic Control of the Emotions;* Matt Ridley (1997), *The Origins of Virtue: Human Instincts and the Evolution of Cooperation;* Richard Wrangham and Dale Peterson (1996), *Demonic Males: Apes and the Origins of Human Violence;* Robert Wright (1994), *The Moral Animal;* Steven Pinker (1997), *How the Mind Works;* Sarah Blaffer Hrdy (1999), *Mother Nature;* and the work of my Emory colleague, Frans De Waal (1996) *Good Natured: The Origins of Right and Wrong in Humans and Other Animals.* A recent book by Pascal Boyer (2001), *Religion Explained,* applies a framework very similar to mine to another area of complex human behavior. (There is even at least one undergraduate text in the field, David M. Buss (1999), *Evolutionary Psychology.*) E. O. Wilson (1998) has recently argued that evolutionary biology will ultimately become the basis for the social sciences.

I write this book in the same spirit. My purpose is to examine, from an evolutionary perspective, certain political behaviors and preferences common to humans. I am concerned in a broad sense with the menu of issues discussed in the political arena. I believe that this menu can be explained as resulting from our biological evolution. Some political issues have been relevant to humans even before they were human, and we have evolved certain behaviors and preferences or ways of looking at these questions. These are the topics that I discuss. Evidence shows that political attitudes are heritable, a requirement for such attitudes to have been selected through evolution (Tesser 1993; Olson, Vernon, and Harris 2001). Evidence of heritability, however, does not imply that there is a gene for each attitude. For example, Olson et al. (2001) indicate that genes may establish general predispositions which may in turn shape environmental influences that may then influence attitudes. As in all behaviors, cause is both genetic and environmental ("nature and nurture"). My point is to analyze the nature part of this causation.

I consider issues at a very broad level because this is the level at which natural selection could have operated. The analogy might be to the evolutionary analysis of language (Pinker 1994), which argues that certain language categories exist in the newborn. As the individual learns the relevant language from their environment, these are filled in. Certain political categories are innate as well, and the details of values preferred for these categories are set as one lives in a society and observes the issues facing that society and the effect of various policies on one's life.

A recent short book by Peter Singer (2000), *A Darwinian Left*, takes an approach somewhat like the one I take here. Singer argues, as do I, that there are evolved political preferences in humans and that political systems must consider, and perhaps adapt to, these preferences. Singer begins with a political agenda, however, and is concerned to show how to best achieve this agenda, given human preferences. I try to be somewhat more analytical and allow the agenda to come from the preferences. That is, to say, I analyze the extent to which political institutions allow humans to fulfill their own preferences, rather than impose my preferences on them. Indeed, in writing this book I have come to question some of my previously held beliefs, as discussed more fully in chapter 6. Nonetheless, I refer to Singer's work at several places in what follows.

We evolved living in groups of perhaps 25 to 150 persons.[1] Political decisions were made in these bands. Now we live in political agglomerations of up to one billion persons, as in China and India. We still make political decisions. The social decision-making mechanisms are radically different today from those in earlier times. But the individual decision-making mechanisms we use are those that evolved with our ancestors, and many of the issues we consider are surprisingly similar to the issues that our ancestors would have found relevant. Sometimes in human history we have done quite a good job of adapting these evolved tendencies in ways that have led to governments that did a good job of satisfying human preferences. At other times, government institutions have been amazingly bad for their citizens and residents. In this book, I show how the constant set of evolved political preferences and behaviors can sometimes lead to good outcomes and sometimes to bad outcomes. I also provide some suggestions for improving our lives today, based on achieving our evolved goals. We can devise institutions that will do a better job of satisfying our evolved preferences, and designing these institutions will be easier and outcomes will be better if we understand the basis for our behaviors and preferences.

I analyze political behaviors in several dimensions: group or social behavior, including ethnic and racial conflict; altruism and cooperation; envy; political power; and the role of religion in politics. I also analyze how we make political decisions. All human societies must deal with these issues—and they have—from the time humans came into being, and even before. As a result, we have evolved a set of preferences and behaviors with respect to these issues. These preferences and behaviors evolved along with our humanness. If we understand the ways in which they evolved, we will be in a better position to understand how we decide on these issues and perhaps even how we should decide.

I reach a surprising conclusion: modern western nations, and particularly the United States, are the most effective societies for satisfying our evolved political preferences. We evolved in small bands of 25 to 150 persons and now live in societies of tens or hundreds of millions. Therefore, it is not obvious that our current society would be good at satisfying our evolved preferences. In this sense, the result is surprising, even remarkable. (The main reason the United States is in a better position than even western European nations is that it does not have any dominant ethnic group, and so the risk of one group seizing the government and using it to impose undesirable predatory ethnic or racial policies is low.)

Perhaps for an economist to write about the biological evolution of political behaviors may seem odd. But modern economics has become an expansive—some say imperialistic—discipline. Economists now study behavior in many contexts and are not confined to the study of only markets. I have done research in public choice, the field of knowledge that applies economic analysis to political questions. I have written extensively about the role of ideology in politics, and it is a natural progression to ask about the ultimate source of that ideology. I have been interested in the evolution of preferences for many years; I have been writing about it since 1979. Other economists (Jack Hirshleifer, Robert Frank, Janet Landa, Gordon Tullock, Sam Bowles, Herbert Gintis, and Arthur Robson, to name only a few) have also written about the interaction between economics and biological evolution. Robson (2001) and Zak and Denzau (2001) have recently written survey articles on the topic. A new journal, the *Journal of Bioeconomics,* is devoted to the issue.

Moreover, many of the behaviors and policies I discuss have economic ramifications. Economists often make policy recommendations. Sometimes policymakers follow these recommendations and sometimes they do not. As a teacher of economics, I see that students often find the lessons of the discipline counterintuitive and often have difficulty understanding what economists have to say—even in areas that economists find uncontroversial. I have always wondered about these issues. Why do well-intentioned policymakers, often following the beliefs of well-intentioned voters, sometimes find our analyses so unconvincing? Why do students find what we economists have to say so surprising, often unacceptable? My first effort to study this issue led me to public choice. In research on voting (often with James Kau of the University of Georgia), I found that ideology is an important determinant of voting. To economists, this result, while now generally accepted, was originally surprising. We expected that economic self-interest would be the main determinant of voting on policies, but it was not.

Ideology was itself not a well-defined topic. There is no good theory of ideology. Political scientists agree that ideology is important, but they do not understand its bases. Psychologists do not have an answer either. I now believe that the basic building blocks of voter ideology come from our evolutionary history. Certain tendencies in political behavior are rooted in this evolutionary history. I hope to elucidate these tendencies in this book, which is an effort at what Edward O. Wilson (1998) has called consilience—the unification of knowledge across the sciences, social sciences, and humanities. I believe that important parts of political behavior are rooted in our evolutionary heritage.

In what follows, I use preferences or tastes in the way economists use the term: some set of desires independent of the costs and prices that determine actual choices exist in humans. These preferences are in part innate and in part shaped by particular elements of culture. The actual outcome of a decision process is the result of the interaction of these preferences with the set of conditions or constraints facing the decisionmaker. This is the basic stuff of standard microeconomic analysis, and I use the term preferences in the conventional way. Psychologists and neuroscientists and others have recently found evidence that the human mind works exactly in this way. Certain factors are linked to the emotions in such a way as to provide pleasure or pain, and these factors are what economists treat as tastes (Damasio 1994; Johnson 1999).

For example: I argue in chapters 3 and 4 that there are preferences regarding altruism. In a political setting, these preferences translate into policies about income redistribution. But culture also shapes desires for altruism. Individuals growing up trained in one religion, for example, may have different preferences than those raised in another tradition. Moreover, even given these preferences, the amount of altruism in a society may vary depending on other conditions, commonly studied by economists and political scientists: How efficient is the tax system? What is the voting system, and how does this effect the political power of the poor? What are levels of literacy in the society? Nonetheless, examining the underlying evolutionary basis for political preferences is still worthwhile—particularly because relatively little literature examines these factors and much examines the more proximate mechanisms.

The standard approach in economics is to take tastes as given from outside. I followed this approach until reading Wilson's sociology in the mid-1970s. I then came to believe that something more could be said about tastes and that biological evolution was the place to find additional information. While all tastes have evolved, some are more interesting than others. For example, we like sugar because our predecessors who ate more sugar were more likely to survive than those who did not and, as a result, became our ancestors. This is a relatively uninteresting taste (unless one is on a weight-reducing diet). On the other hand, natural selection has crafted a rather complex set of preferences with respect to risk. Males take more risks than females, and young males take more risks than old males. These tastes make perfect sense in an evolutionary framework. Tastes for risk are interesting and relevant for many decisions, both financial and political, and are discussed at several points in the book.

I also examine another class of tastes and preferences—those for various political policies. We do not have a deep understanding of why people pursue different political policies. While self-interest can explain some political outcomes, recent research has shown that a very important determinant of voting behavior is ideological. That is, people often do seem to want to vote for the best policy for the general good, but people disagree on what this policy is. I believe that fundamental evolutionary issues can explain these puzzles; this book is an attempt to convince others that this is so.

I take particular positions on many controversial issues found in the literature. I have tried to sort out the controversy and take the position that I believe to be most likely correct or to point out the nature of the controversy and the positions taken in the debate. But one of the most important messages of the book—perhaps the most important message—is that the theory of evolution and the evolutionary history of humans are relevant for understanding contemporary political behavior. Even if the particular details of my argument prove incorrect, the overall perspective should prove useful. Indeed, I cannot conceive of any meaningful alternative; we have evolved, and our politics has evolved with us.

The present analysis will not be able to tell us how a particular society made a particular decision, nor will it be able to directly predict outcomes. This is because the actual outcome depends both on the factors I discuss and other factors, such as the particular culture of a society and the particular circumstances facing a society at a given time. In this sense, it would be completely improper to read my analysis as arguing that biology alone determines outcomes or that I argue for biological determinism. But the evolved factors I identify do have an important influence on final decisions, and the analysis will help us understand one input into that process and the issues facing a political system.

Plan of the Book

Chapter 1 introduces some ideas from biology and the theory of natural selection and some ideas from the history of political philosophy and then shows how these are related. In particular, I argue that the state of nature in which humans were lone individuals and in which there were no rules never existed and could not, in principle, exist. Rather, humans have always had political rules—even before our ancestors were humans. I also explain why desirable policies must treat each person as an individual.

In chapter 2, I introduce the notion of group identity and group conflict. Humans easily form group allegiances and identify with groups. In the evolutionary environment, these groups would have been kin based, but today membership is much more flexible. Humans can learn to be xenophobic, and they can also learn not to be. Ethnic conflict may have been adaptive in the evolutionary environment, but today it is counterproductive for most or all groups. Policies, such as affirmative action, that stress ethnic group identity are dangerous because they can induce people to revert to earlier behavioral patterns.

Chapter 3 discusses the bases for altruism. While humans are somewhat selfish, in many contexts they behave cooperatively. Several theories explain such cooperation. These theories explain why humans have a desire to redistribute some money to the poor and why such redistribution must be coupled with monitoring the behavior of the recipients. I show that utilitarianism as a moral theory is consistent with our evolved preferences; Marxism and the Rawlsian difference principle are not. I also show that our intuitive notions of fairness have a deep basis in evolution.

One human tendency that is important in many policy decisions is envy, which is discussed in chapter 4. Envy has evolutionary bases, rooted in our past as hunter-gatherers. But today envy is probably counterproductive in most cases in western societies. It seems to result from confusing dominance hierarchies with productive hierarchies. Marxism, for example, is clearly based on such confusion. In today's world, unlike the evolutionary world, people mostly become wealthy by being productive and creating benefits for others, and, therefore, desires to punish or penalize the wealthy are misguided.

In chapter 5, I discuss the issue of political power. There is an inherent conflict in interest between rulers and subjects. This conflict has persisted for all human experience, and elements can be found in chimpanzee behavior as well. The nature of our evolved preferences can shed some light on this fundamental political issue.

Chapter 6 analyzes the evolutionary basis of religion. Most humans believe in some religion, and religion has been a major factor in important political behaviors, including wars and massacres and also many cooperative activities. I discuss the evolutionary basis for religion and the cultural selection of religion. Religious beliefs often dictate interference in private behavior. I explain why preferences for such interference may have evolved and discuss their role in today's society.

In chapter 7, I provide an evolutionary analysis of political decision making by individuals. I explain why we often place too much stock in anecdotes and why unusual events are likely to generate too much attention. I also show that many cognitive errors as analyzed by cognitive psychologists Kahneman and Tversky and their disciples, including the behavioral economist Richard Thaler, are the result of asking questions in ways that were not relevant in the evolutionary environment. Therefore, these anomalies may not be as important in explaining actual decision making as might first appear. In this chapter, I also discuss gender differences in political decision making and preferences.

The final chapter is a summary of the policy implications of the analysis for contemporary political decision making. I show how our evolutionary past is still relevant in facing decisions today. Here I show that modern western society does a remarkably good job of satisfying evolved human political preferences.

In writing this book, I did not set out to prove the advantages of the current system in the United States. Indeed, I find many aspects of this system problematic and have criticized it elsewhere. Moreover, when I began this research I was a libertarian and did not understand government regulation of private behavior, such as regulation of drugs or pornography. Based on the reasoning in chapter 6, on the role of religion in human behavior, I now at least understand some social benefits for such regulation. Thus, the analysis has led to the conclusions; I have not used evolutionary analysis to prove points in which I already believed.

Acknowledgments

This book has been in process for several years. It began as a paper, "The State of Nature and the Evolution of Political Preferences," ultimately published in the *American Law and Economics Review* (vol. 3, no. 1, spring 2001). When it became clear that the material was too long for a paper, I began to write this book. As I was writing, numerous people commented on parts of the manuscript. I would like to thank George Benston, Robert Chirinko, Frank Forman, David Geary, Robin Hanson, Bruce Johnsen, Owen Jones, Bruce Knauft, Margaret Levi, Dean Leuck, Richard Epstein, Jack Hirshleifer, Susanne Lohman, Timur Kuran, Richard Posner, Peter Richerson, Frank Salter, Vernon Smith, Som Somanathan, Gordon Tullock, David Sloan Wilson, and Todd Zywicki for comments on various parts of the manuscript. Lee Cronk, Herb Gintis, Michael McGuire, and Joel Mokyr read the entire manuscript and provided very helpful comments. I have also presented parts of the manuscript at seminars at Arizona State University, Emory University, George Mason University, New York University, the University of Montana, the University of California at Los Angeles, and the University of Southern California, and I would like to thank the participants in those seminars. Additionally, parts of the book were presented at meetings of the Association for Politics and the Life Sciences, the Canadian Law and Economics Association, the International Society for Human Ethology, the International Society for New Institutional Economics, and the Society for Evolutionary Analysis in Law. The Gruter Institute for Law and Behavioral Research has been particularly helpful. Emory University has provided a sabbatical for completion of the manuscript. Of course, the work has changed sufficiently so none of those mentioned should be held accountable for anything in the final version.

In addition to the paper in the *American Law and Economics Review*, parts of the manuscript have been published in *Human Nature* ("Hierarchy," vol. 11, no. 3, 2000), the *Journal of Bioeconomics* ("Group Selection and the Limits of Altruism," vol. 2, no. 1, 2000), *Politics and the Life Sciences* ("Does Ethnic Conflict Pay?," vol. 10, no. 1, March 2000), and *Jurimetrics* ("How Humans Make Political Decisions," vol. 41, no. 3, spring 2001), and "Courts and the Tort-Contract Boundary in Product Liability," in *The Fall and Rise of Freedom of Contract*, edited by Frank Buckley, Duke University Press, 1999.

These essays have been rewritten and reorganized, so that no paper can directly be associated with a chapter, but I would like to thank the publishers for permission to reprint the relevant sections.

DARWINIAN POLITICS

CHAPTER 1

Background
Evolution and Politics

The State of Nature

Since at least the work of Hobbes and Locke, the hypothetical state of nature has been an important input into legal, political, and constitutional
analysis. The notion is of humans with no law or government choosing a set
of rules. The question asked is, What sort of rules would such individuals
choose in this situation? A variant, the veil of ignorance, is sometimes used
for policy analysis. The argument here is that the constitutional and other
rules of society should be those that would be chosen by individuals who do
not know their own position in the society that is being designed. People in
this situation are said to be behind the veil of ignorance in that they are uninformed about how rules would effect them as individuals. They would,
therefore, choose rules that would be best for the average citizen.

Several examples of this form of analysis can be found. James Buchanan
(a Nobel Prize laureate in economics) and Gordon Tullock analyzed this
form of thinking of the ideal constitution for a society in *Calculus of Consent* (1965). A second example is from John Rawls's *A Theory of Justice* (1971),
in which he argued that citizens in such a state would choose substantial
amounts of redistribution. (Rawls is discussed in more detail in chapter 4.)
Robert Nozick, in *Anarchy, State, Utopia* (1974), claimed that individuals
would choose a libertarian government. A related question deals with the
primacy of property and of government. It is often argued that the state is
first and that all property rights exist because the state has allowed them
(see, e.g., Sunstein 1993, 4–5). In this view, whatever rights people have are
in an important sense a product of law. A game theoretic approach is found
in Binmore (1998).

Obviously, the state of nature is meant as a metaphor, not as a true statement of primitive conditions. It has the advantage of providing a clean starting point for analysis—what rules and structure would someone starting
with absolutely no law or government choose. Nonetheless, the metaphor is
misleading. That is, the true original state of humans is so different from the
hypothesized state of nature that arguments proceeding from this basis generally cause confusion rather than enlightenment. Similarly, thinking of
property as only defined by law is meaningless; property and law are coequal

and neither precedes the other. In particular, all of these arguments imply that humans have more freedom in choosing institutions and rules than is actually the case.

The argument that in a state of nature there were no rules is analogous to another fallacious argument common in the social sciences—what has been called the Standard Social Science Model (SSSM; Cosmides and Tooby 1992). In the past (and still, in some circles), it was thought that an individual human was (in Locke's phrase) a tabula rasa, a blank slate, and that humans could learn or be trained in almost any way. In this view, people can learn almost anything, and culture can establish almost any rules that we want. Students of human evolution, and in particular evolutionary psychologists, now know that this conception was false and in many respects misleading. In the Preface, I mentioned several recent works that apply this new learning to human behavior. There is room for wide, but not unlimited, variation in culture. Certain individual rules and behaviors are indeed programmed into us, and we violate these rules only at great peril. Similarly, the notion that social rules are arbitrary or that such rules can be purely created by reason is false.

If real policies are based on false constructions, then real suffering may ensue. Pipes (1990) argues that this fallacy, which he traces to Helvétius, a disciple of Locke, was important to the Russian Revolution—an event that did cause a good deal of misery, and whose aftermath is still causing misery. Followers of Marx believed that they could create a new person by changing institutions; they did not believe any inherited nature would make this transformation impossible. That is, the reign of communism in Russia, China, Eastern Europe, and elsewhere was based in part on a faulty analysis of human nature. Marx and his followers believed something equivalent to the Standard Social Science Model and believed that human nature was much more malleable than it actually is. Tullock (1999) suggests that Mao Tse-tung's reading of Engels was a cause of the Great Leap Forward, a policy which killed about thirty million people. It is difficult to imagine any other error in human history that caused more misery. Singer (2000) also discusses Marx's misreading of human nature and indicates that this error "affects much of the thought of the entire left" (5). (I discuss the basis for the Marxist error in chapter 4.)

Evolutionary Basis for Rules

My analysis is directly based on recent work in primate biology. Ultimately, the argument is grounded in sociobiology and evolutionary psychology as applied to humans.[1] The simple point is this: social structure, property rights, and rule-like behavior are older than *Homo sapiens,* so to talk about human beings existing solely as individuals in an environment with no political or legal structure is meaningless. Human beings never existed without rules. By the time our ancestors became human, they already had many rules in place. The simplest and most powerful evidence is from behavior that we observe in chimpanzees, our nearest relative. When two species exhibit similar behavioral or physical characteristics, there are two possible biological explanations. The feature may be analogous—derived and evolved separately—as will be the case when the most recent common ancestor of both species did not exhibit the feature. For example, the wings of birds and of bats are analogous, as are the shapes of fish and dolphins. On the other hand, the characteristic may be homologous—derived through common descent from a common ancestor. If two closely related species exhibit the same characteristic or behavior, and if it appears likely that the ancestral species also exhibited this behavior, it is likely homologous. Moreover, if a feature is homologous in this sense, all intermediate species from the common ancestor to the current species must also have exhibited the same feature.

To the extent that we share traces of political and rule-ordered behavior with chimpanzees, the biological argument is straightforward that the behavior is by common descent. That is, behavior that is common almost certainly is inherited from a common ancestor and did not evolve separately. Just as we can be sure that the common ancestor of humans and chimps (and all ancestral species in between) had five-fingered hands and enjoyed eating sweet foods, we can be sure that this ancestor exhibited certain social behaviors. In particular, for example, chimps, humans, and ancestral and intermediate species lived in groups internally governed by a set of rules and certain hierarchies within those groups and engaged in cooperative behavior within those groups. Moreover, all also engaged in intergroup conflict.[2] In this sense, we may be sure that no human or close ancestor of any human lived a solitary life. Indeed, the books by respected scientists mentioned in the Preface have documented carefully some of the commonalties in social and political behavior between apes and humans. For example, De Waal (1996) has written about desirable features in the behavior of nonhuman apes, such as respect for property rights and enforcement of social rules. Wrangham and Peterson (1996) and Ghiglieri (1999) have written about less praiseworthy aspects of primate behavior, including murder and something very like warfare and genocide. De Waal's work (2001) is a recent compendium of studies by leading primatologists discussing the relevance of primate behavior for human evolution. Ridley (1997) discusses in detail some of the relationships between the biological and economic approaches to the study of humans and human society. An important book dealing with the evolution of group and social behavior is by Sober and Wilson (1998), discussed particularly in chapter 3.

Some apes do live isolated lives. Orangutans live alone, with males only visiting females periodically for procreation and females living with off-spring only until the offspring are old enough to be independent, which occurs at about ten years. Apparently, rape is common among these animals (Wrangham and Peterson 1996, 132–143); of all apes, these may now be the most purely Hobbesian, in that they apparently live solitary lives with almost no social grouping beyond the mother and offspring. However, Dunbar (2001) suggests that orangutans may be more social than they appear, and the apparent isolation in which these animals now live may be due to their living on the margins of their former range. Gibbons live in mated pairs but not in larger groups. The evolutionary history of the apes indicates that gibbons first and then orangutans split off from the ancestor of the remaining apes, so humans are more closely related to gorillas and chimpanzees than to these solitary species. Gorillas and chimps both live in groups, although of different sorts. Thus, the oldest common ancestor that may not have lived in groups was the ancestor of all apes. More likely, however, the ancestral proto-ape did live in groups, since virtually all primates are social (Foley 1995), and gibbons and orangutans have reduced or lost the natural sociality common to most primates.

Maryanski and Turner (1992) argue that apes are less social than monkeys because the ape adaptation of feeding on fruit in trees required less sociality than basic monkey feeding systems. They then argue that this basic adaptation led to a greater desire for autonomy and independence among apes (including humans) than among monkeys. In this view, culture and interpersonal communication among hominids has replaced the innate genetic tendencies of sociality common to monkeys. Even in their view, social tendencies also are quite old among humans.

The argument here is that when our ancestors first became human, they already had in place a set of rules and social institutions. Humans never, at any point in time, lived without social rules. To refer to humans living in a society with no rules or to ask about the behavior of people in such a society is, therefore, meaningless. As our ancestors became more intelligent and more sophisticated and as social groups became larger, the rules would have become more complex, but a state in which humans lived without rules would not be defined. The state of nature would then describe the rules that existed when humans first came into being as humans. To determine what these rules might have looked like, it will be useful to consider what has been called the Environment of Evolutionary Adaptedness (EEA;

Barkow, Cosmides, and Tooby 1992; Crawford 1998; Bowlby 1969 originated the concept). The EEA is mainly the time in which humans evolved from apes or from ancestral species, such as *Homo erectus,* to humans, although some aspects of behavior are even older. Based on these types of arguments, I provide at least a tentative explanation for several aspects of social and political behavior.

The Environment of Evolutionary Adaptedness (EEA)

To understand the applicable evolutionary theory, the characteristics of the environment in which humans evolved must be considered. The relevant time period is the Pleistocene, the period lasting from about 1.6 million to about 10 thousand years before the present (BP), when farming began. During this period, humans evolved from a rather ape-like creature, probably similar to the modern chimpanzee, to modern humans. Intellectually modern humans have probably existed for about fifty thousand years, with the first forty thousand spent as hunter-gatherers (Klein 2000). Selection pressure for certain behaviors and preferences would have led to reproductive success in that environment. The argument of this book is that modern humans have retained some of those preferences and may exhibit some of those behaviors. It is generally believed that the time from when farming began(the Recent or Holocene period) up to the present is too short to have caused significant evolutionary changes in behavior. We share some behaviors with all animals, some with all mammals, and some with all primates, so for these behaviors the EEA is older than the Pleistocene. However, most behavior that is of interest here evolved in the Pleistocene.

Several features of this environment are relevant for understanding the state of nature and current political preferences. The EEA itself would have not been constant, and evidence shows that variability of climate increased during the period of human evolution (Potts 1998). For example, several episodes of glaciation occurred during human evolution. Nonetheless, some attributes probably would have provided constant selection pressure, and these would have had evolutionary implications. Many of these constant features would have had to do with the social environment of our ancestors and, thus, are particularly relevant for analyzing political behavior. It is these relatively constant properties that I discuss. (Potts [1998] argues that the climatic variability of the evolutionary environment means that there were few constant features. But if the main force leading to the evolution of intelligence was competition with other humans, these variable features of the physical environment would not have been crucial.) Features that would have remained constant would have provided continuous selection pressure for behaviors adapted to these features. In general, it would be cheaper for natural selection to seek behavior adapted to these constant attributes than to require individuals each generation to learn how to adapt de novo.

Irons (1998) believes that the concept of the EEA assumes more stability than is appropriate and that some evolution could have occurred since the beginning of agriculture. He also believes, however, that status seeking, an important part of the behavior involved in politics, would have been constant in the evolutionary environment (which he calls the Adaptively Relevant Environment.)

Other evidence of certain constant elements during the course of evolution exists. Wrangham (2001) indicates that the level of sexual dimorphism and other physical characteristics in our ancestors has been constant for 1.9 million years. Therefore, "all humans, from Homo erectus and Neanderthals to ancient and modern Homo sapiens, have lived under the same basic system of social ecology and sexual selection" (143). Bowles and Gintis (2001) examine the ethnographic literature closely and identify eight common features of hunter-gatherers, all having to do with interactions between individuals: groups are small enough so members can observe each other, but large enough for free riding; no centralized governance structure; many unrelated individuals; limited status differences; sharing food and or food acquisition efforts; high membership turnover; no investment; and ostracism as the major form of punishment. Because the important aspects of the environment with respect to political behavior were interactions with other people, the assumption of relative constancy is not harmful. Ambrose (2001) believes that once a certain level of intelligence was reached, further changes were autocatalytic and driven by basically social features.

The method of adaptation would be to generate preferences or tastes for certain behaviors that would have created reproductive fitness with respect to these constant features. Johnson (1999) discusses the emotional mechanisms that served to reinforce preferences for these behaviors. Thus, these constant features would be relevant for understanding current utility functions, the measure of preferences economists use. For example, natural food that tastes sweet always contains sugar, a source of energy. It would have been possible for parents in each generation to teach children to eat sweet foods, but since such foods were always available and always desirable, it was cheaper for natural selection to endow humans (actually, prehuman ancestors) with a taste for sweetness. This book deals with similar tastes for various forms of social interaction.

Social Structure in the EEA

Society consisted of relatively small bands (probably, 25 to 150 beings) of individuals, many of whom were genetically related. Based on our brain size compared with other primates, it appears that we are adapted to groups of 150 (Dunbar 1998). Hominids ("any member of the human family Hominidae, all species of *Australopithecus* and *Homo*" [Jones, Martin, and Pilbeam 1992, 464]) have always been social creatures. During most of this period, bands were probably toward the small end of this distribution and were mobile. At some point, bands probably aggregated into socially significant tribes of up to a few thousand. Within these bands, initially the basic family unit was likely a female and her children and perhaps her mate. As we will see, groups of male kin were also extremely important in understanding human evolution and particularly political behavior. Over evolutionary time, children became increasingly dependent on long-term parental care as intelligence and thus the period of dependency increased.

This increase in dependency that resulted from increasing intelligence would have happened for two reasons. First, increasing brain size at adulthood coupled with a limited-size birth canal means that larger brained (more intelligent) offspring must be born with immature and not fully grown brains. Based on other apes, the human gestation period is actually estimated at twenty-one months, with the last twelve months occurring after birth (Jones et al. 1992, 87). Parents of newborn children will have no trouble in accepting this argument.

Second, increased intelligence means increased value of learning, and so again a longer dependency period. This increase in dependency in turn caused increased reproductive benefits to males derived from spending resources on raising their children, leading to long-term bonding between males and females (Kaplan et al. 2000). Because males who helped raise dependent children would have had more surviving offspring, genes for this behavior would have been selected for and would have increased in the population. Based on relative physical sizes of males and females, humans and their ancestors were likely mildly polygynous (Geary 1998). Some successful males would have had more than one mate. Among polygynous hunter-gatherers, it appears that about 15 percent of males would have had more than one mate.

Ancestral hominid bands made their living by hunting and gathering, with the role of hunting and the importance of meat probably increasing over time. Additional protein available from meat probably indirectly enabled humans to bear the high energetic costs of our large brains and thus led to the increase in intelligence of humans relative to all other species

(Aiello and Wheeler 1995; Foley 1995; Kaplan et al. 2000). Males engaged in hunting, often cooperative hunting, and females in gathering (which explains the preference of women for flowers, as they are associated with things that are good to gather); the sexual division of labor among humans is an evolutionarily old characteristic, probably antedating humanness (Geary 1998; Silverman and Phillips 1998). It is universal among humans (Brown 1991), although it does not occur among other apes. Male and female chimpanzees eat different diets (with males eating more meat), but exchange of foodstuffs is uncommon.

Humans and chimpanzees are each other's closest living relatives; chimps are closer to humans than to gorillas. Chimps and humans are unique even among apes. They are one of the few primate species in which male kin groups are important (Pusey 2001). This means that male bonding within the group is more significant than is true for most other primates and that conflict between male kin groups is more common. Both features are relevant for understanding human evolution and human behavior today. Moreover, humans and chimps are the only primate species in which males engage in coalitional aggression—aggression by an organized group of more than two males.[3] (Tooby and Cosmides, in press; Wrangham and Peterson 1996.) Goodall (1986) and Manson and Wrangham (1991) describe chimpanzee coalitions and aggression in detail. I discuss the implications of these male-based kin groups in both this chapter and in chapter 2.

Natural Selection

Several additional general points regarding natural selection are relevant. The basic mechanism of natural selection is that individuals with certain characteristics are more likely to leave descendants than individuals with different characteristics. Since children resemble parents, the genes that caused the successful behavior are then more common in the next generation. This process, discovered by Darwin,[4] is the theoretical basis for the entire discipline of biology and will ultimately become the basis for the social sciences and humanities as well (Wilson 1998.) Of course, the process is extremely complex, with nuances and complications. The best nontechnical introduction to the theory of natural selection is *The Selfish Gene* (Dawkins 1976/1989). I will introduce additional aspects of the theory of natural selection as they are needed.

Evolution selects for reproductive fitness—propagation of genes into the future. As subjects of natural selection, we need not understand its workings. It works whether we understand it or not, and whether we try to help it or not. Moreover, natural selection works with whatever tools are available. The simplest and most basic tools are the primitive emotions that are hard wired in the primitive parts of our brains. These emotions are then coupled to behaviors and preferences that lead to reproductive success. (This link is described in detail in Johnson 1999.) No one (human or animal) needs to understand that sugar provides physiological energy in a convenient and easily digested form. All that is required is that sugar tastes good and that eating sweet foods is pleasurable. Our predecessors who had this set of preferences would have sought out sweet food, accumulated energy, and outbred those who preferred other less nutritious foods. Of course, this preference for sweets evolved long before humans were humans, but our ancestors retained it, as do we.

We can make a similar argument about sex. Natural selection had no need to program a desire for children or for propagating the species. Rather, humans and other animals desire sex, and this reliably led to offspring in the circumstances of the EEA. Natural selection has programmed in most cases among mammals, including of course humans, a desire to care for children once they are born, and this, coupled with the desire for sex, has been enough to propagate the human species and all other surviving species as well. We enjoy those things that caused our ancestors to survive and leave more offspring. Contemporaries of our ancestors with other tastes did not become our ancestors—either because they themselves died too young or because they did not have offspring to leave us their genes.

Natural selection is itself a great economizer. It will use whatever mechanisms are easiest and most readily available. If two factors are usually associated and one is easier to identify, then natural selection will use the easiest as a clue to behavior. If incest is biologically undesirable because children of incestuous relationships are less fit (as is true) and if individuals raised together are generally siblings, then a method for siblings to recognize each other in order to avoid sexual relations is not needed. Rather, humans and other animals have evolved to find sexual relations with those with whom they were raised undesirable. While this mechanism is not perfect (sometimes unrelated individuals will be raised together, as in Israeli kibbutzim, and sometimes siblings will be raised apart), it was relatively easy to evolve and good enough to work. To evolve an actual kin detection mechanism would have been difficult, if not impossible. In general, in the EEA humans lived in relatively small groups and many group members were kin. Therefore, certain behaviors with respect to group members evolved; natural selection did not need to teach us much about how to recognize kin. (This point is discussed further in chapter 2.) We will see similar mechanisms in other aspects of behavior.

In the evolutionary environment, polygyny and limited contraception were common or universal. Therefore, behaviors that would have led males to acquire additional mates would have also led to increased offspring. Irons

(1998) points out that status (defined in whatever way is relevant for the particular society) would have had this effect. Consequently, males who sought and achieved status would have had more offspring than other males. Since the connection between status and reproductive success was universal, there was no need to determine if this connection was present in a particular instance and no need for learning. The result is a very common desire by males for status and success, even in current environments where such success is divorced from increased fitness.

The form of preferences (what economists call utility functions) would be that form that would have led to maximum fitness, or reproductive success. That relationship between preferences and consumption, or wealth, would have been selected and would have led to maximum reproductive fitness in the EEA. For example, in some circumstances a certain amount of altruism—of charitable or sympathetic behavior—would have been fitness maximizing. On the other hand, in some circumstances, relative rather than absolute wealth may be the relevant variable, which explains envy, discussed in chapter 4. And risk preferences (Rubin and Paul 1979; Rode and Wang 2000) and time preference (Rogers 1994) can also be explained in terms of evolutionary evolved tastes. Additional examples are provided throughout this book.

Individuality

Different humans have different preferences, so it might appear that evolution cannot explain these preferences. But there are evolutionary explanations as to why different people might have different preferences. One important point is that males and females pursue different reproductive strategies, and so behave differently. This difference in behaviors would then have been caused by different preferences. (I discuss these gender differences in political preferences in chapter 7.) For now, it is useful to understand that males have always been more involved in political behavior than females, so when I use the masculine pronoun to refer to actors in the political arena in evolutionary times, my choice is based on the actual facts of the evolutionary situation.

But even among males or among females, differences in political preferences are evident, in part because of different individuals being in different situations. All else equal, a poor person would be in favor of more income redistribution than would a wealthy person if only because the wealthy person would pay for the transfers to the poor person. Farmers will favor farm price supports, professors will favor subsidies to education, and textile workers, tariffs on textiles for the same reason. This is not the complete answer, however. Even aside from narrow self-interest, true differences between people in beliefs about what society should look like. That is, political preferences and ideologies differ fundamentally; we do not all have the same set of preferences, and we do not all view the good society in the same terms.

One important explanation for these differences is that natural selection will not make us all identical. Rather, fundamental to the process of natural selection are differences in evolved tastes. The process itself will generate systematic differences between individuals. Various mechanisms lead to such differences.

One such mechanism is that in many circumstances the return to various actions or strategies depends on the number of other agents who play alternative strategies. In this case, the return to a given strategy is said to be frequency dependent, and if this is so, then the evolutionary outcome will entail different strategies in the same population. This notion is a result of a branch of knowledge called evolutionary game theory, first expounded in John Maynard Smith, *Evolution and the Theory of Games* (1982) (for a recent exposition, including elements of both biology and economics, see Gintis 2000a).

The best known example of this situation is in the so-called hawk-dove game. Assume that there is some resource (say, a piece of fruit or, in some circumstances, a fertile female) and two individuals reach this resource simultaneously. Each can play one of two strategies: hawk or dove. A hawk will fight for the resource; a dove will give in to a hawk and search elsewhere. Then if a hawk meets a dove, the hawk wins. If a dove meets another dove, the winner is chosen randomly. If two hawks meet, they will fight and the victor will get the resource. But fighting is costly on average, perhaps for the winner as well as the loser because either or both may be injured or even killed. As a result, the return to a dove meeting another dove is greater than the return to a hawk meeting another hawk, even though a hawk has a higher return than a dove in a hawk-dove meeting. The result is that population equilibrium results from a mixture of hawks and doves. To see this, think of beginning with a population of doves. If hawk mutant then arises, the mutant will do very well because most of his meetings will be with doves that will give in during conflicts. This is an evolutionary game, so the successful hawk will leave a larger number of offspring in the next generation than will the average dove; the proportion of hawks will grow over time. But as the number of hawks grows, the chance of two hawks meeting and fighting will increase. At some point, the number of hawks meeting each other will be sufficiently large so that the returns from being a dove will on average just equal the returns from being a hawk. At that point, the population is in equilibrium with some hawks and some doves. This mixture of hawks and doves is said to be an Evolutionarily Stable Strategy (ESS) and it implies two types of individuals in the population. (One interesting variant is to assume that one individual gets to the resource first. Then a third strategy, called bourgeois, can arise. This is a strategy of "hawk when you are first, dove when you are second." This strategy will dominate both hawk and dove, but it requires some asymmetry in the positions of the players, and so may not evolve.)

The story can be generalized. Think of two strategies in a cooperative business endeavor (or, from evolutionary times, a cooperative hunting endeavor): be honest, or cheat. Individuals are chosen randomly to be partners, and each can be honest or cheat. Two honest players earn more together than two cheaters, but a cheater earns more than does an honest player if they become partners. Then if we begin with all honest players, the percentage of cheaters will increase in the population. At some point, cheaters become sufficiently common so that they are meeting and cheating each other sufficiently often so that being a cheater pays exactly the same return as being honest; at that point, both types exist in the population and there is equilibrium, an ESS. This game has some other interesting characteristics. In particular, each individual will want to pair only with an honest individual. Therefore, both cheaters and honest individuals will try to appear honest, but individuals will also develop mechanisms for detecting cheaters who are trying to appear to be honest. This is an arms race between cheating and detection of cheating. (We come back to this point in chapter 3, where we will see that the cheating detection mechanism plays an important role in understanding political behavior.)

Another issue is the extent to which individuals have preferences for fairness or are self-interested, as assumed in most of neoclassical economics. A good deal of evidence reveals that people are not as self-interested as economists have traditionally assumed. This evidence is both experimental (e.g, Fehr and Schmidt 2002) and based on real world observation (e.g., Ostrom 2000). I will not go into details, but models of evolution of such behaviors indicate that the population will include both selfish and non-selfish types (e.g., Sethi and Somanathan 1996; Gintis 2000b; Bowles and Gintis 2001). Experimental evidence shows that the percentage of each type differs across societies and depends in part on economic conditions in the society, but all societies have both types (Henrich et al. 2001). There appear to be at least three types: selfish; cooperative; and those willing to incur costs to punish the selfish individuals.

In all cases, the equilibrium result is that both types of individuals—hawks and doves, honest players and cheaters, selfish and cooperative—will exist in the population. The actual percentage of each in a given population will depend on the various costs and benefits of each strategy, but the nature of the outcome is clear—some of each type will occur in the population. Just from these three examples we have generated eight types of individuals—cooperative honest and cheating doves, cooperative honest and cheating hawks, and similarly for selfish individuals. If we realize that many such decisions and games may be found in a real population and each may have more than two equilibrium strategies, then it is easy to see that there will be many types of people. Additionally, in many cases, strategies will be mixed. A mixed strategy involves randomization of play, so opponents cannot predict which strategy any given opponent will play in a particular encounter. Thus, even people with identical preferences and strategy sets may choose different strategies simply because of randomization. Moreover, strategies in general will be continuous, not discrete. Few people are always honest and few always cheat. But thresholds of honesty or cheating can vary and probably vary along a continuum. Thus, even on one dimension, many possible strategies are available. As we add dimensions, the number of potential personalities increases quickly.

A similar analysis of frequency dependent selection may explain sociopaths, individuals who are self-centered, impulsive, and unable to form personal commitments. These individuals, who compose 3 to 4 percent of the male and less than 1 percent of the female population, commit a large percentage of crimes in a population. Mealey (1995) uses a frequency dependent evolutionary game theoretic model to explain this behavior. In her model, individuals who are socially disadvantaged are more likely to cheat. Extreme cheaters are sociopaths. Under some circumstances, such behavior will pay, so there is an equilibrium number of sociopaths in the population. On the other hand, a society composed entirely of sociopaths, which would be like a prison with no guards, would not be viable.

It is useful to note that, although game theoretic notions are useful in explaining human behavior, there is no presumption that we actually think in the terms used by game theorists. Boyer (2001) points out that humans engage in social interactions but do not understand these interactions. He also indicates that describing behavior in terms of strategies, utilities, signaling, and other game theoretic concepts strikes people as alien. But this does not mean that the concepts are not useful; it merely indicates that they are not accessible to conscious reasoning.

Other mechanisms lead to individual differences. An example is shown by sickle-cell anemia. (This discussion is based on Wills 1998.) This is a genetic condition in which red blood cells are defective, and the disease can lead to early death; before treatment was widely available, death rates were 15 percent in the first year of life. This genetic disease is caused by having two genes for sickling. (Having two identical genes at one place is called being homozygous.) But having one gene for sickling and one not (being heterozygous) is quite useful in an environment where malaria is common. Thus, at equilibrium three types are found in the population: those with no sickle cell genes; those with one; and those with two. Geneticists call this a balanced polymorphism. In West Africa, where malaria is common, in some tribes up to one-half of the population carries one gene for sickling, with the incidence of sickle cell anemia then being up to 25 percent. Although sickle cell anemia is the best known malaria-related balanced polymorphism, Wills indicates that there are several others, including thalessemia among Mediterranean populations (especially in Greece, Italy, and Sardinia). Thus, where conditions are proper for the benefits of such balanced polymorphisms, several types of individuals will exist in the population. The case of balanced polymorphisms has been most carefully studied for malaria-related genes because of the deleterious effects of the gene in America where malaria is not a threat. It is likely that balanced polymorphisms are much more common and that many other instances, including behavioral examples, will be found.

These are examples of reasons why evolution will not generate the same sets of genes or preferences in everyone. The result is that humans are highly individualistic. We differ from each other on numerous dimensions. Reiss (2000) has indicated sixteen dimensions on which humans differ, of which at least fourteen are genetically based, and Arnhart (1998) has identified twenty. Differences on this number dimensions can generate huge numbers of differences between individuals; Reiss (2000) identifies two trillion different possible profiles. This individuality is added to the desire for independence discussed by Maryanski and Turner (1992). It also explains why human individuality is important, and why political ideologies or theories that assume everyone to be the same are doomed to failure. Humans have different genetic endowments and have different tastes, leading them to choose different strategies, which is an argument for heterogeneity in society. These differences explain why individuals want some freedom from social control.

Preferences and Behavior

Evolution has endowed us with certain preferences, those that would have been fitness maximizing in the EEA. However, humans are extremely facultative—that is, able to change behavior in response to environmental changes. Humans are the most facultative animals in the world. Indeed, this facultative behavior is the basis of economics, which studies behavioral responses to changes in relative prices. (The nature of facultative behaviors as understood by biologists is discussed in Sober and Wilson 1998.) Thus, while we may have certain preferences based on evolution, we are able to modify behavior in response to changes in environments—relative prices, where price is broadly defined to include all costs and benefits associated with a choice. These responses are the subject matter of economics. The arguments in this book are about preferences, or utility functions. Economists understand human behavior as the maximization of preferences subject to constraints, including prices. Thus, actual behavior is only partially determined by preferences. To explain observed behavior, prices or other constraints must be considered as well.

For example, think about the taste for sweets and fats. This taste evolved in the EEA at a time when starvation was a continual danger, and it was necessary to eat food that was as nourishing as possible whenever such food was available. In particular, fats and sweets provided useful high quality energy, and those who became our ancestors were the members of the population who were best able to find and use this energy. But today, we understand that these foods may be harmful in too great quantities (may have too high a price in terms of health problems), and many people make great efforts not to consume them. (For an interesting discussion, see Burnham and Phelan 2000.) Food and chemical companies also invest large sums in creating foods that can fool our tastes, and many of us eagerly buy these foods. Thus, though there is a taste for such foods, the taste is only one input into the decision process, and other elements are involved in the actual decision as to what is consumed. Philipson and Posner (1999) have presented an explicit economic model of the effect of changes in technology and thus in relative prices on the amount of obesity in a society.

The entire discipline of economics studies exactly how people respond to changes in prices, given tastes. The most basic law in economics, the law of demand, tells us that people change their consumption in response to changes in prices of goods. In this analysis, price is to be interpreted broadly. Part of a price may be the time needed to consume some good. Prices would also include health effects and other nonpecuniary implications of consumption. The maintained hypothesis in economics is that tastes are constant, but from this assumption and the assumption of approximately rational behavior, a major system of analysis is possible. The contribution of biology is to provide a basis for these tastes; the basis is of course fitness in the EEA.

A biological example is sexual behavior, the heart of fitness. We all have certain preferences for sex. But behaviors have costs. Traditionally, the cost was an increased risk of pregnancy. The invention of various forms of efficient birth control and the availability of abortion has reduced the cost of sex, and people engage in more of it than was true even a generation ago. The perceived risk of diseases and particularly AIDS for a time increased perceived prices, and people responded to that price as well. The existence of welfare programs reduced the price of children, and recent modifications in welfare programs may have increased those prices. In each case, sexual behavior responded though basic preferences remained unchanged. Thus, on an underlying unchanging set of preferences, changes in all sorts of costs have induced changes in behaviors, even in the most fundamental biological behavior. Economics studies these changes in prices. But economics has always lacked a theory of tastes; this is the role of evolutionary psychology in the analysis. Here I develop the beginnings of an analysis of underlying political tastes.

Additionally, I generally ignore the role of culture. But culture also has an important effect on actual behavior and operates through both prices and preferences. For example, the changes in sexual behavior discussed have changed the cultural perception of premarital sex (by reducing the cost in terms of social disapproval) and so have contributed to the trend toward increased sexual relations of unmarried people. Moreover, culture and genes can evolve together, and some of the evolutionary processes I describe may have coevolved with culture. (This does not apply to the sexual behaviors discussed earlier—at least not yet.) Boyd and Richerson are leading advocates of this view (see their publications cited in the bibliography). In a major work on the relation between genes and culture, Durham (1991) defines his task as trying to explain human diversity across cultures. In this book, I concentrate on commonalties among all human societies. That is, I am concerned with tastes that humans share. In any given culture, the way in which these tastes express themselves varies, but the set of tastes remains constant. In this sense I can largely ignore the role of culture.

Humans can learn. Learning interacts in complex ways with preferences. First, preferences govern (Damasio 1994). To think of behavior without preferences is meaningless. We behave in order to satisfy preferences. Thus, preferences come before learning, or any other type of behavior.

Second, however, we can learn the price of satisfying preferences, and the extent to which we satisfy any given preference is a function of price. This is the subject matter of economics. The price may be a conventional money price, but it can be more general. Part of the price of eating sugar and fat is reduced health, and people respond to that price as well. I show in many places in this book that the price of certain political preferences is too high, and people may refrain from satisfying those preferences as the price increases. For example, we may have something of a preference for ethnic conflict, but as the price increases, we can learn not to indulge this taste.

Finally, some things are more easily learned than others. If some situation were common in the evolutionary environment, natural selection would have equipped us to respond to that situation in a fitness maximizing way with little or no learning. Much sexual behavior is of this sort as is some political behavior, such as status seeking. Other situations are relatively novel with respect to evolution, and learning how to deal with these situations may require more explicit learning. For example, everyone easily learns to talk, but must be taught to read and write. If a situation is sufficiently novel, it may be useful to learn how to satisfy the preferences that govern that situation. One point of this book is to demonstrate how certain political behaviors may be counterproductive with respect to our evolved preferences in the novel environments in which we now live. We can then learn that satisfying these preferences costs too much, and decide not to satisfy them.

Thus, it would be incorrect and counterproductive to view the analysis in this book as explaining the actual outcome of behavior. It would also be incorrect to view arguments in this book as implying that behavior is solely determined by the factors identified here. I do not argue in any sense that political behavior is biologically determined. Rather, factors considered in this analysis are only one input into the determination of final outcomes. However, they are an important input, and understanding these factors will improve our understanding of human behavior. If we understand exactly what political tastes natural selection has provided to us, we can better understand political behavior. Moreover, learning that some tastes may not be productive in the current environment (in the same way that a taste for fat is not productive in an environment where all the fat we might want is around the corner at the local grocery store) may be helpful. We may then be in a better position to modify our behavior and improve outcomes.

A Zero-Sum Society?

One economic point is important in understanding the evolved nature of political preferences. In the EEA, there probably were very few gains from trade, except for exchange between men and women. That is, resources and incomes were probably relatively fixed, with little possibility of value increasing exchange or production, such as we see all around us today. Therefore, we may not be well adapted to think intuitively in terms of gains from trade. This explains why the results of the economic analysis of trade—that all parties gain from trade, and that free international trade is welfare maximizing—are counterintuitive. Of course, just because certain forms of explanation are counterintuitive does not mean that these explanatory principles cannot be learned. Economists have learned to understand these results at a deep level, and the increasing worldwide reduction in tariffs may mean that we have succeeded in teaching them to others. But these results must be studied and taught; they are not learned intuitively.

The result is that humans in many cases now tend to base decisions on zero-sum thinking when other forms of analysis would be useful. The term zero-sum comes from game theory and refers to a game such as poker, where the gains and losses of all the players add to zero. Other situations may be positive sum, where the gains of all players are greater than the losses. This characterizes exchange, where everyone can benefit. Negative-sum games are where the total losses are greater than the total gains, such as occurs in war. Many examples of cases where zero-sum thinking will occur are found in this book.

Before considering the zero sum nature of the EEA, some preliminaries are in order. We must distinguish between specialization, division of labor, and exchange. Division of labor occurs whenever individuals are not self-sufficient. Among humans, division of labor by gender is universal (Brown 1991). Whether other patterns of division of labor existed in evolutionary times is not clear. Ridley (1997, 49) believes that such division of labor was common; Maynard Smith and Szathmary (1999, 148) believe that it was uncommon until relatively recent times—"Populations of, at the most, a few hundred individuals, with little division of labour, except, probably, that between the sexes, have been replaced by societies of many millions, dependent on extensive division of labour."

In modern societies, specialization, division of labor, and exchange are closely related. Individuals specialize in tasks and are compensated either through directly exchanging their output for other goods and services in markets, or by being paid by those who have engaged in such exchange. Moreover, complex tasks are broken into parts and each individual carries out only a small part of the task; this is division of labor, which adds greatly to productivity. Adam Smith, in the classic *Wealth of Nations* (1789/1994), described in detail the workings of a pin factory. He showed that the number of pins per worker was tremendously increased (in Smith's example, from less than 20 to 4,800 hundred per worker per day) through the division of labor and specialization. Again, compensation occurs when the outputs of the process (the pins) are is sold in a market and workers are paid wages.

However, division of labor and exchange can occur without specialization, and this may have been the pattern in the EEA. Exchange might be intertemporal exchange of the same good, as when a successful hunter shares his prey in the expectation that when he is less lucky he will be the beneficiary of a similar deal. This is the insurance function of exchange in primitive societies discussed by Knauft (1991) and many others. It is often argued that reciprocal altruism (Trivers 1971, discussed in chapter 3) might apply to intertemporal exchanges of the same good of this sort. Similarly, division of labor might occur without specialization. Consider a hunting process in which a group of hunters surround the prey that is to be killed by that hunter in whose direction the animal runs. Here each participant is performing the same task (so there is no specialization except a trivial geo-graphic specialization), but there is division of labor. The hunters will then share the game.

Apparently, there was relatively little division of labor other than by gen-der in evolutionary times. Of course, gender-based division of labor was very important and itself greatly increased human productivity (see, e.g., Knauft 2000). But there are several reasons for believing that more general division of labor was relatively unimportant. First, group sizes were relatively small, and as Adam Smith first indicated in 1776, "The division of labor is limited by the extent of the market" (1789/1994, 19). Before a pin factory can ex-ist, sufficient demand for pins to support it must exist. Among small groups, there was not enough room for substantial division of labor. If a social group is small, then a specialist will not be able to work full time at the specialty be-cause the group will not be able to make use of the output of a full-time spe-cialist. For example, if a flint knapper can make twenty hand axes per day and if his band only uses twenty axes per week, then the knapper will only work at flint knapping one day per week. As Carneiro (2000, 12929) indi-cates, "full-time craft specialists come into being only when the aggregate demand for their products has reached a certain threshold." Competence at a specialty increases with the amount of time spent on the specialty. Therefore, if someone is less than a full-time specialist, he will not develop as high a degree of competence as would be possible with more time spent on the specialty. (If the knapper worked at it every day, his output might in-crease to twenty-five axes per day.) This will in turn reduce the benefits of specialization and may explain the sad situation among the Tasmanians. Edgerton (1992) describes these people, who lived in bands of forty to fifty people within a total population of about four thousand. The Tasmanians at one time had a relatively advanced technology developed before the flooding of the land bridge that had connected them to Australia. Over a ten thousand year period of isolation, they progressively lost most of their technology. This may simply be because the size of the population and of the small bands was insufficient to support any significant level of division of labor. Edgerton describes other societies that lost various forms of tech-nology as a result of isolation or other sources of population reduction.

In discussing his vision of the division of labor, Ridley (1997, 49) says: "One man made stone tools, another knew how to find game, a third was es-pecially good at throwing spears, a fourth could be relied upon as a strate-gist." But in a group of 50 to 150 individuals, the likely group size during much of the EEA, there would not be full-time work for most of these spe-cialties. Rather, while people may have had the skills Ridley mentions, they would have been unlikely to engage in these activities on a full-time basis. That is, while there may have been some limited amount of specialization, it would have been incomplete. The hunter who knew how to find game would have thrown spears at it as well, and the spear thrower might have engaged in butchering if his spear hit.

Sahlins (1972) also argues that the division of labor and specialization was limited in primitive societies of the sort that are relevant for analysis of the EEA. He quotes Marshall from her 1961 book, "for every man can and does make the things that men make and every woman the things that women make" (1972, 9). While Sahlins's book is in general controversial, it appears that this particular point has not been debated. For example, Bird-David (1992), in a well-known paper commenting on Sahlins, discusses the simple nature of the tools used in such societies, with the implication that specialized toolmakers did not exist. Keeley (1996, 46) makes the same point about warriors: "societies without specialization in the economic realm were unlikely to develop specialized warriors or units."

Moreover, exchange would have been difficult, and so again limited. Economists stress that in nonmonetary societies, barter would have been necessary for exchange. A requirement for double coincidence of wants limits barter—that is, each party to a potential exchange must have exactly what the other party desires for exchange to occur. Reciprocal altruism extends this possibility somewhat by allowing intertemporal exchange, but the amount of nonsimultaneous exchange is still limited by the possibility of cheating. That is, if one hunter shares his game today, he must be certain that the beneficiary will share his kill when situations are reversed, but guaranteeing this in a world with no contract enforcement is difficult. The biological notion of reciprocal altruism is a mechanism that alleviates but does not fully solve this problem. This limited amount of exchange is consistent with the conclusion of Stiner et al. (1998) that "low human population densities during most of the Middle Paleolithic imply that group sizes and social networks were small, which certainly limited the numeric scope of individual interactions. Under these conditions the possibilities for evolution of complex sharing and exchange behavior as ways to counter the effects of unpredictable resource supplies would have also been quite limited." This suggests that even the insurance function of exchange was limited.

The key distinction seems to be between simple and complex hunter-gatherers (Knauft 1991; Kelly 1995.) Simple hunter-gatherers are mobile, egalitarian, live in small settlements, and the only occupational specialization is by age (Kelly 1995, table 8-1, 294.) Complex hunter-gatherers differ in many dimensions; in particular, occupational specialization is common. Thus, this distinction (which Kelly attributes to changing from a mobile to a sedentary life style) is the border between specialized and unspecialized roles for individuals in societies. By the time of large agricultural societies, division of labor was universal and important among humans. But this came late in human prehistory—probably too late to have left a significant mark on our evolved preferences.

The result is that we evolved in a world with limited possibilities of gains from trade or exchange and limited possibilities of activities that would increase personal or social wealth. Therefore, our minds are built for understanding a zero-sum society. Economics stresses the possibilities of gains from trade—it is possible for both parties to come out ahead from exchange and this is what runs a modern economy. This notion is counterintuitive for most people, even many educated people. We will see the operation of this point at many places in the analysis.

Indeed, even when people do engage in mutually beneficial trade, this mutual benefit does not motivate them. Rather, each aims at maximizing his or her own benefits. The fact that for trade to occur, both must benefit is irrelevant for each individual. To quote Adam Smith again, "It is not from the benevolence of the butcher, the brewer, or the baker that we expect our dinner, but from their regard to their own interest" (15). Thus, there is no reason to expect that an innate mental module to measure gains from trade has evolved. Rather, we are each selected to try to be sure that we gain from trade; gains to our trading partner are irrelevant. Moreover, mental mechanisms work against this recognition of mutual benefit. Even in mutually beneficial trades, an aspect of competition is found. Both sides want to engross for themselves as much as possible of the gains. That is, buyers want the price to be low, and sellers want it to be high. Therefore, in engaging in trade, an important consideration is to avoid being victimized. As a result, mental modules aimed at policing transactions have evolved (Cosmides and Tooby 1992). These modules focus on the zero-sum aspect of trade (Wright 1999)—that aspect dealing with the terms of the bargain, rather than with the gains.

There may be another difference between conflict and trade. To engage in conflict (at least at the group or tribal level), a conscious policy decision must be made. Many persons must be consulted, and pros and cons debated. On the other hand, individuals automatically engage in mutually advantageous trade. This might mean that we must think more consciously about gains or losses from conflict, while we can simply engage in trade without any conscious thought, and especially without any thought of the mutual benefits.

I do not want to be read as implying that trade is artificial or that people must be taught to engage in exchange. As an economist, I believe that trade is of fundamental importance to humans; Adam Smith discussed "the propensity to truck, barter, and exchange one thing for another" (14). People engage in mutually profitable exchange when it is possible, and they do so automatically. But for reasons discussed above, I do believe that people do not have an innate understanding of the benefits of such exchange; if they did, the teaching of economics would be much simpler. This understanding is an example of something that can be learned but that is not innate.

This lack of understanding may apply particularly to international trade —trade between countries or, more relevant for the analysis here, between ethnic or tribal groups. In addition to the normal views of trade, international trade also implicates our xenophobic modules. Krugman (1997) points out the fallacies in many highly popular arguments against such trade; the relevant point for the analysis here is that the anti-trade arguments were highly popular and presented in well-selling books. Such arguments are also popular politically. While the United States now has relatively small tariffs and other barriers to international trade, there was a struggle to reduce these levels, and there is a continual struggle to maintain these low levels, in part because people do not fully understand that trade is mutually beneficial. Even today, in debates about trade, people focus on jobs (for which trade is irrelevant) rather than the income improving aspect of trade. The antiglobalization movement, which manifests itself in protests against various aspects of international trade, may be a further example of similar preferences.

The consumer movement, associated with Ralph Nader, also focuses on the competitive aspects of exchange. Exchanges between buyers and sellers, or consumers and businesses, are mutually beneficial. But the consumer movement focuses on the terms of the transaction, rather than on the mutual gains, and tries to shift the terms. In some circumstances, the result is actually reduced trade and fewer transactions, so consumers may actually lose from this set of policies. For example, when consumerists succeed in passing laws requiring higher quality but more expensive goods than some consumers might prefer, some potential buyers are priced out of the market and end up as net losers. The focus on the terms of the transaction is an evolved tendency from our zero-sum past.

It may also be true that for much evolutionary time, there were limited returns even to greater productive effort. This would especially apply to hunting (Tudge 1998.) If a hunter increased his efforts, the amount of game in the nearby area would be reduced, and the costs of future hunting increased. In such a setting, there would have been little incentive to increase efforts and little ability to learn that greater effort leads to increased rewards.

It is also worth noting that possibilities for productive investment were also limited in the EEA. Tools were used, but only small tools in limited numbers. Because societies were mobile, creating more goods than were easily portable did not pay. Similarly, investing in large durable structures such as houses did not pay, because these would be abandoned when the next move occurred. As a result, we may not have intuitions about the productivity of capital or investment. This may explain why some religions have forbidden charging interest, as discussed in chapter 6. It may also be related to certain errors in Marxist analysis.

Kin

I now examine the role of kin in behavior. Interestingly, this was not understood until 1964, when William Hamilton wrote two extremely important papers in the *Journal of Theoretical Biology*. The basic argument is this: we share genes with our relatives, so if a gene leads to behavior that improves the fitness of relatives, it also increases its own chance for survival. Parents and children, and full siblings, are related by a factor of one-half; half-siblings and grandchildren are related by one-fourth; first cousins by one-eighth; and other relatives are related by lower percentages. This means that the probability of any given gene of an individual being found by common descent in the individual's parents or siblings is 50 percent. So if a gene for altruistic (helping) behavior should arise and lead to an individual helping a sibling, there is a 50 percent probability that the gene will also benefit itself. If some form of assistance is more productive (worth more) to a sibling by a factor of two relative to its value to the individual, it is in the interest of the altruistic gene to induce the individual to provide that aid.

For example, if one's brother had an unlucky hunt and will starve without some additional transfer, it is one's genetic interest to provide him with food, even aside from any expectation of receiving the equivalent amount of food in return at a later date. This is because saving one's brother's life by a relatively small transfer of meat increases his fitness by more than twice as much as the reduction in meat reduces the successful hunter's fitness. Of course, some kin benefits are obvious to anyone: parents spend large amounts of resources on their children. But Hamilton's point was much more general; many kinds of activities can be supported by kin selection, and it does not apply only to parental care.

Kin have many genes in common. Kin selection is said to exist when an altruistic behavior (which would include cooperation in a prisoner's dilemma, discussed in chapter 3) toward relatives is selected because this increases the representation of the shared gene. (For a discussion, see Dawkins 1976/1989). To the extent that members of groups are related, this mechanism comes into play to explain cooperation. Altruism toward kin is dependent on costs and benefits: as the cost to the altruist decreases or the benefit to the recipient increases, and as relatedness increases, such altruism becomes more likely. An altruistic act pays as long as $c < rb$ where c is the cost of the altruistic act, r is the degree of genetic relatedness between the actor and the beneficiary and *b* is the benefit to the recipient of the altruistic act. (This inequality is called Hamilton's rule.) In other words, if some act costs me fifty units of fitness but provides my brother or child (related to me by one-half) with 110 units of fitness, then this act is worthwhile. Genes that induced animals to undertake such altruistic acts would have increased in the population over time, and we should be selected to undertake such acts. Kin based altruism coupled with an unusual inheritance mechanism is the explanation for highly developed cooperation in many insects, such as bees and ants as discussed in E. O. Wilson's *Sociobiology: The New Synthesis* (1975), the first book written on sociobiology.

It is because of kin selection that the human and chimpanzee breeding and residence systems are so important. Both humans (during the EEA) and chimps are said to be patrilocal; that is, male kin stay together and females migrate upon reaching maturity. (Migration by members of one gender or the other in social species is the mechanism that avoids incest. Among most social primates, females remain in the birth troop and males migrate.) Therefore, in any given community groups of related males—fathers and sons, brothers, cousins, uncles, and nephews are found. Because of this relatedness, cooperation among related males is more likely. A payoff from cheating on some cooperative arrangement that would tempt a stranger would not be worthwhile for a relative, because some of the cost will be borne by an individual with whom the cheater shares genes. As a result, groups of male kin can undertake activities that are not available to unrelated individuals. These activities might be cooperative hunting of relatively more dangerous game, a productive but risky activity where shirking is possible. But they may also be attacking the tribe in the next clearing. As we will see, this latter type of predatory activity has been quite important to humans and their ancestors, and it remains an important activity of many. In an important paper, Bingham (1999) has stressed the significance in human evolution of the increasing ability of our ancestors to punish cheaters and thus control shirking.

Recently, Wolfe (2000) has argued on biological grounds that Americans are hostile toward the inheritance tax for reasons consistent with the arguments here. He points out that a desire to leave resources to children is a form of kin based altruism, and people can understand this altruism. According to Wolfe, this explains why a majority favors repeal of this tax, even though most will not directly benefit.

Males and Females

This is a good place to discuss general biological differences between males and females. One important difference among humans and their relatives has just been discussed: humans have been patrilocal, so males in a band or group in evolutionary times were in general related. Males could more easily form coalitions, and coalitions are the essence of politics among humans and even among chimps, as shown by Frans de Waal. Thus, even if females had the desire to be active in politics in the EEA, they would have been handicapped by the lack of close genetic relatives with whom to form coalitions. But females would not have a strong desire to be politically active anyway.

Although patrilocality is relatively rare in the animal kingdom, other differences between males and females are almost universal, at least among mammals. Males invest less in children than do females, and a male is able to sire many more children than any one female can bear. The implications of these simple facts are far reaching.

First, polygyny (some males having more than one mate) has been common in most human societies. (The opposite, polyandry, females having more than one mate, has been extremely rare. When it exists, the multiple husbands are generally brothers, as would be consistent with kin selection.) Dominant males have had more than one mate and have fathered more than their share of children. The reproductive payoff to a male of becoming dominant has been large, and so it has paid (genetically) for males to invest resources in attempting to become dominant. A male becoming dominant has been able to increase his fitness by much more than would have been true for a female, because females in the EEA generally had the maximum biologically feasible number of children anyway. Those males who succeeded in becoming dominant have left more genes in our population than males who were not dominant, so we have more of those genes and some desire to become dominant ourselves. But among humans and our relatives, one way to become dominant is through politics and formation of coalitions. This is another important reason why politics has been a male's game, since patrtilocality has meant that males would have had more genetic relatives with whom to form coalitions.

Dominance contests are not merely through formation of coalitions. There are actual physical contests as well. Homicide is always and everywhere disproportionately a man's game, whether as perpetrator or as victim. Moreover, many homicides are related to dominance struggles, and often involve what seem to be trivial issues. But this point is not the issue at risk; it is the goal of winning and becoming dominant and being in a better position to obtain access to females. (Daly and Wilson 1988, is an excellent study of homicide from an evolutionary perspective.)

Second, males in most species take more risks than females. If a male takes some risk and loses, he may not sire any children. But if he wins, he may sire two or three (or even more) times as many children as a male who does not take the risk. If a female takes some risk and loses, then again she may not bear any children. But if she wins, she cannot increase her fertility by much if at all. The number of children any woman can bear is given by physiological constraints, which would have been binding in the EEA. As a result, the genes of risk-taking males have been passed on to us, and males even today are more risk seeking than females. As Kingsley Browne (1998) has pointed out, this difference in risk preference and risk-taking behavior effects us in several ways. Women commonly take less risky jobs than men do (where risk may be defined as financial risk or as physical risk). Males also take more physical risks, and males are hurt or killed in accidents at a much higher rate than women. Males drown at a rate four times the rate of women, are killed or injured in all sorts of play more often than females, and are burned more often than females (Geary 1998, 319). Any parent of young boys and girls or any observer of a playground will not find these results surprising.

Third, even among males, differences in risk behavior are apparent. In particular, young males will take greater risks than will older males (Rubin and Paul 1979). This is because if a young male does not succeed in breeding, from the perspective of natural selection, he may as well die, since in either case he will leave no genes. Therefore, in evolutionary times, those young males who did not take risks and did not breed also did not leave any genes for us today. We are descended from those young males who took risks and won, and we bear their genes and behave as the genes would suggest. The higher automobile insurance fees paid by young male drivers is only one manifestation of this difference in contemporary behavior.

Again, I want to reiterate that the behavior I refer to is embedded in tastes; there is no conscious calculation. Males who speculate on the foreign exchange market and hit it big, or males who enter politics, need not consider the effect of their behavior on the number of offspring they will sire and need not consider number of children or even sexual access as a goal. (I will at this point not insert a tasteless joke about a recent president.) But in the EEA, males who undertook the equivalent behaviors would have sired more children, and those males were disproportionately our ancestors. Therefore, the tastes that led to that behavior are represented in our genes, and we share the tastes and behaviors, if not the outcomes.

Modules

Evolutionary psychologists stress that the mind is not a general purpose machine. Rather, it is a series of modules, each specialized for a particular task. (This is the message of Barkow et al. 1992, the most important book in evolutionary psychology.) The reason for modularization is essentially economic: for a general purpose mind to learn all that is necessary to function in the world would simply be too costly. Thus, particular modules evolved to handle problems that were common in the EEA. Of course, the mind can now handle many other problems, but these are often treated through learned mechanisms. Many of the implications for psychology are discussed in Barkow et al. (1992) and in the journals *Human Nature* and *Evolution and Human Behavior*. The origin of the theory was from studies of language, where it was determined that no one could learn everything that is necessary to speak a language de novo (Pinker 1994). I will indicate instances where it appears that some political variable is related to a particular module, but this will not be an important or crucial aspect of my analysis.

Societies, Ancient and Modern

One important difference between the EEA and modern societies that has ramifications in many contexts is the difference in scale in the societies. Societies in the EEA were small—perhaps 25 to 150 persons, with occasional larger aggregations for purposes of exchange and mate seeking. I live in the United States, a society of about 280,000,000. While many of the groups I interact with directly are of the scale of groups in the EEA (e.g., my department has about fifteen full-time members), many are vastly larger (the American Economics Association has about twenty six thousand members; Atlanta, Georgia, my home, has about 4 million people.) This difference in scale has many implications. Many of the people I depend upon for necessities (the farmer who grows my food, the firms that process it, the grocer who sells it) are complete strangers, and I will never meet them. Many of them live in other countries or continents—the Chinese or Indian seamstresses who sewed my shirts, the farmers who grew the cotton. We see in chapter 3 that mechanisms ensuring cooperation in small environments where interactions are face-to-face are quite different from mechanisms needed in large rather impersonal environments such as our contemporary world. And we see in chapter 7 that some methods of decision making carried over from the EEA are often not well suited to decision making in larger environments.

This should not in any way be interpreted as arguing that we suffer from living in large groups or from living in environments different from the EEA. This book does not argue that we are worse off or that returning to environments more like those of the EEA would benefit us. On the contrary, I argue that we still have tastes and preferences that evolved in the EEA, but that we are today better able to satisfy these tastes than were our ancestors. In other words, those of us fortunate to live in modern Western societies (and perhaps most living humans) are happier than most of our ancestors.

We have their tastes, but much higher real incomes and much better technology to satisfy these tastes.

I have mentioned human tastes for various types of foods. But in the EEA and throughout much of recorded history, merely getting enough food to survive was often a problem. Now we have enough (for many, more than enough) food. Moreover, it is easily and readily available in tasty and nutritious forms. We have year-round access to seasonal products such as fruits and vegetables. We like the foods that were nutritious in the EEA, but we have much better access than our ancestors and at much lower real costs; for example, in the United States in 1998, only 14 percent of total personal consumption expenditures are on food. In the EEA, most human effort was aimed at food production—at hunting and gathering.

Humans are curious, as befits a species that has advanced based on intelligence. In the EEA, storytelling and mythmaking were methods of satisfying human curiosity. But today virtually everyone in America and other advanced western societies has access to sources of information and entertainment that would have been unimaginable to our ancestors (even to our grandparents). Television brings dozens of channels to our home, with entertainment and information on each one. Every city has libraries and bookstores. Although the technology is new, the Internet is an amazing source of information. Those of us for whom curiosity is a driving force can specialize in informational occupations, such as becoming professors, where we spend all of our time seeking and disseminating information.

Bands in the EEA were relatively egalitarian with respect to adult men. After the EEA, with the rise of agricultural societies, extreme inegalitarianism became the norm, and kings and other dominant males in early societies engrossed great power and numerous women for themselves. They were also vastly richer than were others. We have now returned closely (not fully) to the egalitarian norms of the early societies. In particular, monogamy means that with respect to the major biological element of fitness, we are again egalitarian; almost all men who desire to do so can obtain a wife and have children. (These issues are discussed in chapter 5.) What inequality exists generally occurs because some individuals are vastly more productive than others or are descended from such individuals, but this productivity benefits all in society, as discussed in chapter 4. Indeed, one of the themes of this book is that overall, political institutions in the west are better adapted to fulfill human preferences than has been true of any other human society. (Maryanski and Turner 1992 and Kronk 1999 also argue that modern societies are more consistent with human evolved preferences than were agricultural societies.)

Human societies have never been particularly egalitarian with respect to the status of women. Even in the EEA, in the relatively egalitarian societies that existed then, equality was mainly between men. In early historical societies, women were also poorly treated. Modern western societies are more favorable toward women than any other societies that have existed. Women have different political preferences than men (as discussed in chapter 7), but the same preferences with respect to equal treatment (Hrdy 1999). Modern society satisfies that preference better than has any other culture, which is an important reason for my argument that current society is preferable to others.

As discussed in chapter 2, war and intergroup conflict has been a major component of human life. For individuals in modern western societies, war is less of a threat than has ever been true in the past. Even if we include modern wars, death from violence among modern humans is lower than at other times in the past, including even hunter-gatherer societies, where death from violence is remarkably common. If we had the death rate from violence that characterized societies in the EEA, then in the twentieth century, twenty times more persons would have died from violence related sources than did in fact die from such causes (Keeley 1996). While there are remnants of intergroup conflict and discrimination in western societies, these are remnants. Members of minority groups are probably better treated in the modern United States than has been true of any other society, and the level of intergroup conflict and prejudice is remarkably low.

Finally, we live longer and healthier lives. While many individuals worry about the dangers of contemporary life, the truth is that our life expectancies are longer than has ever been true and are increasing. Our ancestors would have liked to have lived longer (especially if they could have maintained their health), but the technology was not available. Thus, in this area as well, we are vastly better able to satisfy evolved preferences than has ever been true in the past.

We are economically better off now than in the past. Additionally, we are politically better off. That is, modern western society does a better job of satisfying our political preferences than has any other. (This point is made most strongly in chapter 8.)

Summary

The state of nature, a situation in which humans had no rules, is not a useful metaphor for studying human political behavior because such a state never existed. Humans were never humans with no rules. Political philosophies such as Marxism that assume complete malleability of humans are misguided and often very harmful.

To understand our evolved behavior it is necessary to consider the environment of evolutionary adaptedness. This was a world of relatively small bands of humans or proto-humans, with bands of perhaps 25 to 150 persons who made their living by hunting (mostly men) and gathering (mainly women.) Intelligence increased over the time of the EEA, as did pair bonding between males and females.

Humans and their closest relatives, chimpanzees, are patrilocal—groups of related males live together. This increases the possibility of male bonding and cooperation for activities such as hunting and predation against neighbors. Although all humans have been selected by natural selection, there are important forces for individual differences as well. Natural selection chooses preferences, but humans are very flexible in the ways we react to these preferences in response to changes in prices, broadly defined.

In the EEA, there would have been relatively little division of labor (except gender-based division of labor, which is universal among humans) and few gains from trade. This is because the small size of human groups limited the possibilities of specialization and gains from trade and may mean that we have evolved intuitions consistent with a zero-sum society. There was also relatively little investment, and so we do not have intuitions regarding the productivity of capital and investment. All of these intuitions can be changed through teaching.

A male can father more children than a female can bear. This ability increases the payoff to males from activities that generate access to more females. The result is that males are much more risk taking and much more political than females. These behaviors were also enhanced by the patrilocal system of humans, meaning that males were related to their neighbors, increasing the payoff from group-based activities, including war and politics. Most political actors in the EEA, and even today, are males. Males who engaged in such activities were disproportionately likely to be our ancestors, so we have inherited tastes for such activities. These tastes can persist even if they are decoupled from the fitness returns that would have been common in the EEA. Overall, modern western society satisfies human desires, which evolved in the EEA, better than has any other society.

CHAPTER 2

Groups

Membership and
Conflict

Group Membership

A key distinction that all humans make is between members of the group and outsiders (Brown 1991, 136). The group identification mechanism in humans works easily and almost automatically. We are able to identify with almost any group, no matter how arbitrarily defined. To see an example, just go to any sporting event. A team of young males (or, occasionally, and only recently, females) of various ethnic backgrounds who are from all over the country—sometimes all over the world—and who have no particular personal connection with their team's home city play another similar group of young men in American football, basketball, soccer, or baseball. The stands are full of cheering fans, many of whom may be migrants from other cities, or even other countries, who identify with our team, which may even have been located in another city last year. This is a surprising behavior from many standpoints, but it seems to be a human universal. Call a team the local team, and citizens immediately begin to identify with it and cheer for it.

The behavior becomes more understandable if we include two evolutionary facts. First, the actions of the players are closely related to what would have been military actions in the evolutionary environment. Running, throwing projectiles (balls), kicking, hitting with clubs (bats, hockey sticks), and knocking down opponents—all of these actions are direct modifications of ancestral actions that would have been related to defense from others or offense against them. Second, in the evolutionary environment, the lives of our ancestors often depended on the strength and prowess of their young males. If these young males were more effective than those of competing tribes or clans, these predecessors survived and became our ancestors; if they were not, we are not descended from those individuals. Even the behavior of the fans—for example, shouting instructions to the players—could be descended from evolutionarily useful behavior, as older, more experienced but less strong males would have been teachers of the young males.

Sports teams and identification with these teams is only one example of the group identification mechanism common to humans. (For an early discussion, see Tiger 1969.) Indeed, even infants distinguish between group members and nonmembers. Among infants, males make this distinction more strongly than females, and this gender-based difference carries over to adults as well. Moreover, although common physical appearance makes identifying group members easier for infants, it is not necessary (Premack and Premack 1994). All humans also recognize subgroups based on kinship, sex, and age (Brown 1991, 137). Humans quickly classify people as members or nonmembers of the in-group. Members and nonmembers are treated very differently. It appears that relations between group members and others are handled by different mental modules. All nonmembers are viewed as being alike and as being enemies, while members are treated as individuals (Krebs and Denton 1997).

Groups have certain features in common: Members behave so as to provide benefits to members but not outsiders and expect such benefits from other group members. Members of one group associate behavior of members of other groups with the group itself (the Green Team cheats) rather than with a particular person (Tom Jones cheats). Humans view groups as being purposive, even though groups as groups cannot have any goals. Members of groups, and particularly males, put pressure on other members to conform to group behaviors and accept group beliefs or ideologies. Members are also concerned with the loyalty of other members and desire to punish defectors, or even those who fail to punish defectors (Boyer 2001, citing Kurzban 1999).

Although humans distinguish between members and nonmembers of a group, the group can be almost anything. Group affiliation mechanisms are quite flexible, and the definition of the group is quite variable. Creating in-groups is easy. Social psychologists have shown that people identify with groups based on small or even artificial differences. For example, in one experiment, when students were randomly assigned to a group, they immediately identified with this group, even though they knew that the basis for assignment was random. Social psychologists routinely use such mechanisms as asking students to estimate the number of dots flashed on a screen and then arbitrarily identify the subjects as underestimators or overestimators, supposedly based on their responses. Subjects then identify themselves as members of the relevant group (Sidanius, Pratto, and Mitchell 1994). Tajfel (1970) and Mullen, Brown, and Smith (1992) discuss a large number of studies from social psychology on the importance of ingroup bias, which is generally substantial.

Rubin and Somanathan (1998) discuss the implications this flexibility has for employee productivity and find that productivity is greatly improved in situations where employees can work effectively with members of different ethnic or cultural groups. In this example, being an employee of Acme

Widget Co. defines the group. Kuran (1998) shows that strength of ethnic identification is variable and can shift very quickly. Sober and Wilson (1998, 92) define a group in biological or fitness terms: "A group is defined as a set of individuals that [sic] influence each other's fitness with respect to a certain trait but not the fitness of those outside the group." This definition is consistent with individuals belonging simultaneously to more than one group, with different groups associated with different traits. Boyer (2001) argues that the mind uses the group identification mechanism to identify members of a coalition and to enforce nondefection within the coalition.

In the EEA, male members of the local group or band would often have been fairly close relatives because males have been patrilocal, as discussed in chapter 1. Therefore, kin selection would have been important in creating group identity and in inducing cooperation among group members. When large groups or bands in hunter-gatherer or other simple societies split into two separate groups because of population pressure, the split is along kinship lines, with each new village or group having the more closely related members from the original clan. This means that members each of the new groups are more closely related to each other than to members of the other daughter group.

This close genetic relationship in bands in the EEA may explain why humans seem to have evolved to be quite flexible in their group preferences. If everyone nearby were a relative, a mechanism distinguishing kin from unrelated neighbors would not be needed. Crawford (1991) suggests that methods for identifying closer and more distant kin might been based on any one of a number of criteria: geographic propinquity, treating nearby individuals as kin; treating frequent associates as kin; treating individuals physically similar to oneself as kin; and treating individuals with particular genetic markers as kin. Similarly, Shaw and Wong (1989) suggest that group markers in addition to being in the same band or groups might include language, religion, phenotype (i.e., body type and appearance), homeland, and myth of common descent. Group identification is stronger as more of these markers are comparable. In the EEA, these factors would have been associated with being kin, but such commonalties today can exist independently of actual relatedness. We are no longer particularly patrilocal, and neighbors and frequent associates are not especially likely to be kin for most of us.

This argument is analogous to the argument regarding incest avoidance as being monitored by avoidance of sexual relations with those people with whom one was raised. In determining group identity, genetic relationship is inferred from clues that in the EEA would have been associated with such relationship. In both cases, natural selection focused on characteristics that are easy to monitor and closely correlated with actual kinship, which is much more difficult (perhaps impossible before DNA testing) to measure or directly observe. In a conflict with a neighboring group, all members of one's group could be viewed as kin. Even if everyone in the band were relatives, it would still have been useful to identify closer and more distant kin, and the same clues might have been used. All of these mechanisms can also lead to identifying with unrelated individuals in modern environments.

Indeed, some organizations exploit this inability of humans to determine which individuals are genetically related and to use other cues to identify kin. The Mafia is famous for organizing itself on artificial kin lines and calling itself a family. College fraternities and sororities have brothers and sisters. Religions treat members as brothers. Nations also use family-based metaphors (motherland or fatherland) to benefit from the kin selection mechanisms evolved in the EEA.

In addition to the experimental evidence cited earlier, the types of individuals with whom modern humans feel solidarity show the flexibility of whatever kin or in-group recognition mechanisms actually operate. Humans today identify with groups based on actual kin or ethnic groups but also many other factors. All humans live in families, but all also view themselves as members of larger groups as well (Brown 1991). Some examples are: members of the same religious group; citizens of the same country or (with respect at least to sports events) the same city; graduates of the same high school or university (again, often for sports events); members of the same occupation or profession; employees of the same division within a company (or department within a university) or of the same company; members of fraternal groups, such as the Lions; members of the same gender. The key distinction is between rules involving individuals in the same group and rules involving outsiders; all humans make this distinction.

Moreover, we can be members of many groups simultaneously. I myself am an *American,* a *male,* a *member of a family,* a *professor* who is an *economist* of *Jewish* descent teaching at *Emory University* with *Libertarian* political tendencies and living in *Atlanta, Georgia.* Depending on circumstances, I might feel loyalty to groups defined over any of these characteristics—and I do not consider myself a joiner. Moreover, each of these identifications is itself flexible. I am a member of a nuclear family, but for various purposes my family identification can be extended to include aunts and uncles, nieces and nephews, cousins, grandchildren, my wife's relatives, and perhaps others. Similarly, identifications can be narrowed: I am an economist, but a microeconomist specializing in law and economics, public choice, and bioeconomics. After the September 11, 2001, attack on the World Trade Center, Americans increased the strength of their identification as Americans, as shown, for example, by the increased display of the American flag.

Group identity is hierarchical. This is a feature of many primate groups

(Dunbar 1998). We may be a member of one group that is itself a subset of another group. For an example, consider my profession of academic economist. Within economics, I may identify myself as a microeconomist when the department is considering whether to hire a microeconomist or a macroeconomist. But when the college is considering whether to award a position to a sociologist or an economist, all economists join in decrying sociology. Economists and sociologists may agree to hire a social scientist instead of a humanities professor. Humanities professors and natural and social scientists will all agree that the budget for the College of Arts and Sciences should be large relative to the professional schools, but we agree with our colleagues in the Medical, Law, and Business Schools that the university needs more money. In a company, an employee who is in the widget-marketing department in Indianapolis may identify with that department of his company for some purposes. He may sometimes support the entire marketing department in opposition to production or engineering, and sometimes the entire Indianapolis region instead of the San Francisco or Seattle office. In other cases, he may support the entire widget line against other products made by the same company. In still other contexts, he may defend his company against other firms, and for yet other purposes, he may argue for the benefits of private enterprise against government regulation, or for domestic manufacturers as opposed to foreign producers and importers of widgets. Moreover, if he changes jobs tomorrow, his loyalties can change almost immediately.

This sort of flexible hierarchical group membership is a natural and universal aspect of human behavior, and the ability to form such affiliations is quite valuable. In the EEA, it might have led to survival in conflicts between groups. Those with more flexible membership rules would have been more likely to win such competitions. This flexibility is not a logically necessary part of human identity, and we could conceive of a species like us in many dimensions but lacking this flexibility. Indeed, some argue that Neanderthals lacked this flexibility, based on the lower population density of this closely related species (Richerson and Boyd 1999). But once the genetic ability for such flexibility arose, it would have been quickly chosen by natural selection. Without it, the large societies in which we now live would be impossible.

Religion can serve to define members of the relevant group. All human groups have some religious beliefs (Brown 1991; Burkert 1996; Boyer 2001). In one major survey of beliefs, 77 percent of the respondents indicated that they believed in God, and in no country did fewer than 36 percent believe in God (Inglehart, Basañez, and Moreno 1998).[1] I discuss later individual benefits from such belief. It is likely that religions were originally tribal, so coreligionists were also members of the same ethnic and kin group. The major successful world religions (Christianity, with 33.5 percent of the world's population and Islam, with 18.2 percent[2]), however, seem to have overcome this ethnic basis and to have grown by inducing individuals to accept unrelated coreligionists as members of the same group. For example, Stark (1997, 213) points out that one reason for the rise of Christianity was that it offered "a coherent culture that was entirely stripped of ethnicity." Of course, religions tend to divide on what appear to outsiders to be arbitrary theological distinctions. While many wars have been fought and are still being fought over religious differences, I conjecture that the inclusiveness of the modern major religions has net reduced warfare by implicitly expanding the size of groups. Keeley (1996) does indicate that rates of death from war in primitive societies are much higher—up to twenty times higher—than in modern societies. (Other roles for religion in the EEA and today and the basis for religious belief are discussed in chapter 6.)

The fact that the human group membership mechanism is so flexible is a major biological characteristic of humans. It is easy to conceive of humans as existing without this characteristic. That is, as a species, humans might have been as intelligent as the actual human species but much more constrained in group flexibility. Such a species would live in small kin-based groups. This is how our hunter-gatherer ancestors lived for most of our existence. This social structure would eliminate many benefits of the large agglomerations in which humans actually live. For example, the benefits of the massive division of labor in modern societies would not exist. I believe that this flexibility of group membership is an important biological fact that is ignored because it is taken for granted.

Ridley (1997, 168–169) makes the point that we could conceive of humans as being less tribal and living in more open societies; the extent of our identification with the group is a variable. Ghiglieri (1999) also argues that humans are ethnocentric and xenophobic. The point here is the opposite: we could also be more groupish (Ridley's term) than we actually are. As mentioned earlier, Neanderthals may not have been able to form large tribal groups, as large as those formed by modern humans. If so, this may be one explanation for the success of modern humans and the failure of Neanderthals. For modern humans, group identity is relatively open biologically, although social institutions may constrain it. We could easily be less open and more racist and tribal than we are in fact. If we were more tribal, then multiethnic countries such as the Roman Empire and the contemporary United States could not exist. At one time our ancestors may have been more xenophobic, but as the benefits of increasing group size (successful predation against smaller groups and avoiding being victims of predation by larger groups) increased, there was evolutionary pressure toward more openness.

Often it is asserted that humans are by nature xenophobic and hostile to outsiders (e.g., Shaw and Wong 1989; Ghiglieri 1999.) Of course, some humans do behave in this way, as discussed later in this chapter. But the level of such hostility is facultative—a behavior that can change in response to different environments, including cultural environments. That is, group identification is responsive to changes in relative prices, where price is defined in broad terms. As racism or xenophobia becomes more expensive or has greater costs, fewer people will adopt such positions, as is true in contemporary America. As racism becomes more accepted, as was true in Nazi Germany, more people will become racist. Kuran (1998) indicates that ethnic identity may have a tipping point as more people in a group become ethnically active. The point is that in fact our group membership selection mechanisms are actually quite flexible, and it would be easy to think of humans as being much more clannish and xenophobic than we are. If this were so, we would live in a very different world but would still be recognizably human.

This flexibility would have had advantages in terms of competition with other groups. Abilities to switch membership from one group to another would have meant that in the event of a war or battle, a member of a losing band would be able to join another group rather than being killed or dying of starvation. Keeley (1996) points out that in primitive warfare, captured men are almost always killed, though females may be incorporated into the victorious group. However, he also indicates that many tribes have been eliminated, with survivors joining other tribes or groups. Thus, individuals with such flexible membership mechanisms would have been more likely to survive and reproduce. Sober and Wilson (1998) also indicate that unsuccessful groups or tribes commonly dissolve and members then join more successful groups. Today, emigration allows humans to leave less successful environments and move to places where they expect to be more successful. The ability to join groups arranged hierarchically would also have been useful in a conflict situation.

Through evolutionary time, one force would have been the increasing size of groups. If one set of hominids were able to coalesce into a larger group made up from several subgroups, then this group would have an advantage in conflict. Survival of rivals would be enhanced if rival groups could do the same. Alexander (1987) indicates that these balance of power considerations could lead to any size society. Diamond (1998) indicates that a major increase in group size occurred as a result of farming. Carneiro (2000) says that it was not until humans became sedentary and population density increased that hierarchical military groups were formed. This idea is consistent with Alexander's arguments: defense and predation created a demand for or value of larger groups, but this demand could not be met until agriculture enabled larger groups to feed themselves. On the other hand, once a battle was won, the ability to garner as much of the spoils for oneself and one's closer kin at the expense of nonrelated or more distantly related group members would also have been advantageous. This is exactly the sort of behavior with respect to members of the in-group that is observed.

Conflict

Social living has substantial costs, and a species will be social only if benefits outweigh these costs. For example, conspecifics (members of the same species) are competitors for resources. Also, conspecifics are rivals for mates, generating internal conflict. Social living enables disease to spread more efficiently; contagious diseases, major killers in historic (and presumably prehistoric) times, cannot exist until population density reaches a certain level, as discussed in Diamond (1998, chap. 11). Until recently, disease killed more soldiers than did actual warfare (Keeley 1996), which indicates that the benefits even of larger military groupings were limited by this factor. Larger groups must travel farther and more often to find sufficient resources or must protect larger territories. Thus, living in groups generates large costs of various sorts. Therefore, in many circumstances, it would seem to make individual sense to live a relatively solitary life.

One of the major benefits of group living is the ability to minimize predation. There are economies of scale in this activity—as groups become larger, the ability to avoid predation increases more quickly than group size. Group members can warn each other of the approach of a predator, thus reducing the average time that each individual must spend in defensive observation. They can also unite to mob or fight off a predator that is larger or more powerful than any individual. Most group-living species do so for exactly this reason. In the case of humans, there is no nonhuman predator sufficiently powerful to explain the large size of human groups (even in prehistoric or evolutionary times). Humans did quite well in the EEA with small groups. But predation from other human or hominid groups is and has been sufficiently powerful to induce humans to live in increasingly large groups.

Even among chimpanzees, in the documented cases of genocide against neighboring bands, the larger band has always won. That is, the force leading to larger group size has been the danger from being attacked by other groups of humans or proto-humans. At any given time, a larger group has an advantage over a smaller group, so groups are under continual pressure to become larger. Indeed, this process may be continuing. These pressures were not directly the result of evolution, but because humans have evolved with a tendency to predate on others, the advantage of large group size is ultimately due to biology. Much of written history is an examination of forces leading to increasing group size, and much is a discussion of competition between groups.

There are also diseconomies of scale as groups become larger, such as the difficulty of coordinating the activities of a larger population. This is the essence of politics and is discussed more fully in chapter 5. Indeed, brain size correlates strongly with group size across many species (Dunbar 1998). Another limiting force has been the difficulty of obtaining food for a larger group. The advent of agriculture meant that this limit became less restraining, and, therefore, food production became less of a constraint. At any time and for any state of technology, the balance of factors may determine the size of groups. The point here is that predation always provided one force for increasing group size, but other forces may have countered this factor at any given time in history.

Competition and Intelligence

The most remarkable feature of humans is our level of intelligence, which is sufficiently high so as to be qualitatively different from any other species. Evolution of this level of intelligence is an extremely improbable event in biology; it has occurred only one time in all the billions of years of evolution of life on earth (Mayr 1988). For such a level of intelligence to have evolved, a positive feedback process would be necessary. The only process that fulfills this requirement is the value of intelligence for competition with other intelligent entities—that is, with other hominids. This has been called the Machiavellian hypothesis or the social intelligence hypothesis.[3] Such competition or combat has the needed properties. As one band or group (or species) became more intelligent, its competitors must also have become more intelligent or have lost out and been killed. A key point for understanding the evolution of human behavior and intelligence is that conflict has long been an important part of ape behavior (Manson and Wrangham 1991; Wrangham and Peterson 1996) and was undoubtedly important in human evolution as well.

Competition and predation would have provided powerful pressures for selection for intelligence from several sources. Groups with more intelligent members would have had an advantage in conflict with groups of less intelligent individuals. One benefit would have been better tactics and perhaps weapons. As group size became larger, more intelligence would have been needed simply to keep track of individual relationships with other group members. Dunbar (1996) argues that speech evolved to replace grooming behavior as a method of forming bonds with and monitoring behavior of individuals in groups of increasing size. He claims that speech became necessary as group size increased beyond the level that grooming could support. One form this monitoring of individuals would have taken would have been monitoring members of one's own group to detect shirking in combat. After all, in a combat situation, there is great danger and a strong incentive to shirk, or free ride, as discussed in chapter 3. Indeed, Tooby and Cosmides (in press) argue that many animals could benefit from forming aggressive coalitions but are not intelligent enough to avoid the cheating that such coalitions would engender.

Chimpanzees and humans, because of their male kin bonding system, place more emphasis on competition between groups than other apes (Foley 1995). Groups of kin can more easily overcome various free rider problems than nonrelated individuals. (For a position that questions this argument, see Stanford and Allen 1991.) As mentioned in chapter 1, humans and chimps are the only apes and two of the few species engaging in competition between large coalitions. This is apparently because such coalitional competition requires greater intelligence to monitor behavior and control free riding than is available to other species. Predation and intergroup conflict was so pervasive in proto-human history that the selection pressure from this conflict generated our massive level of intelligence. While data on prehuman hominids is lacking, Keeley (1996) does indicate that warfare in primitive human societies is a more significant source of death than in advanced societies, even when major wars are included. This pressure for evolution of intelligence driven by competition can be self-reinforcing and can lead to a positive feedback system and runaway competition. In this argument, formation of kin-based coalitions was a crucial step in triggering the process.

Competition within the group for resources and mates would have become more intense, and, therefore, selective pressure within the group for increased intelligence would have intensified, as discussed by Miller (2000). Between-group pressures would have selected intelligence in males, who are the main fighters. However, within-group pressures would have selected intelligence in both males and females. Because males invest significant resources in children, females would have sought more intelligent mates who could provide more resources. But exactly because males also invest in children, males would also have sought more fit females for long-term mating, where fitness would include intelligence, so that children in whom the males invested would have been more fit. For short-term liaisons (which males are also selected to seek; see Buss 1999), fitness of females would have been less important since males would not have invested any resources in offspring of such unions. Females, on the other hand, would seek unions with higher quality or more intelligent males, even if they were not their long-term mates, to improve the quality and fitness of their offspring. (For a very helpful discussion of the evolution of gender differences, see Geary 1998. This point is stressed in Miller 2000. Political implications are discussed in chapter 7.)

Rose (1998) presents an interesting variant of this argument. In his version, the precipitating process was tool use. Tools made proto-humans more effective at killing each other and created an incentive for calculated rather then genetically programmed strategies. This then created an additional value to intelligence. But this extra intelligence at some point becomes free, because it is also useful for hunting or other food and resource acquisition activities. This could explain the rapid growth in human intelligence. At some point, the value of intelligence for food acquisition is maximized, and the final push is through interspecies competition—competition with other proto-humans. One interesting implication of this story is that selection for intelligence would be for a general-purpose intelligence, useful in more than one domain.

One key force in selection within the group would be selection to engage in opportunistic cheating and to avoid being the victim of a successful cheater. Cosmides and Tooby (1992) have shown that the human mind has evolved to be well adapted to detect cheating. They have also shown that humans are well adapted to distinguish fair weather from true friends (Tooby and Cosmides 1996). The ability to punish such cheating is at the heart of Bingham's (1999) theory. The other force would be the ability to function as a member of a hierarchical group including both relatives and nonrelatives. This ability would have been necessary to compete successfully with other groups, where competition would have included actual combat. Tooby and Cosmides (in press) discuss in detail the complex cognitive mechanisms necessary for successful intergroup competition, including the ability to monitor numerous coalition members to assess contributions and measure risk bearing. They argue that only a few species have been sufficiently intelligent for this behavior, though individuals in other species would benefit from such behavior if they were able to perform it.

Thus, to understand the state of nature, we must replace the Hobbesian world of individuals in conflict with a world of groups in conflict. This difference in perspective has profound implications for the nature of the evolution of rules. A world of individuals and conflict between individuals is truly Hobbesian; initially there would be no rules, and the world would be one of each for himself. To reduce conflict, these individuals would eventually have created rules. In a world of groups and conflicts between groups, behavior within the group would have been governed by existing, evolved (not created) rules. If we allow for group selection (Sober and Wilson 1998), then those groups that used the best rules would be the most successful at competing with neighboring groups and would, therefore, expand their share of the population. Even without this mechanism, private mechanisms for selecting more efficient group rules would have been in place, as discussed in chapter 3. Thus, rules governing social actions of individuals would have come into being along with humans themselves. Asking about the life of human beings living in isolation with no social structure would not be meaningful. Moreover, since rules evolved along with humans, asking what rules humans in a totally ruleless state would choose is also meaningless. Such a world has never existed and, in principle, cannot exist.

Supply and Demand for Coalition Size

This is a place where economic principles can again help improve our understanding. Coalitional competition can lead to an increased value or demand for intelligence. However, for this demand to actually lead to increased intelligence requires additionally a "supply" or resource side. Initially, for the growth of the brain, supply was in the form of sufficient sources of energy from meat to support a larger brain, which is a calorically or energetically expensive organ. Meat also enabled our ancestors to reduce the size of their digestive system, again leading to an energy savings (Aiello and Wheeler 1995). Much later, agriculture became the supply force leading to further increased size of human groups. We may identify three equilibria: Chimpanzees and our early ancestors ate relatively little meat, and the equilibrium was the small band, held together by ancient primate social forces. Increased meat eating led to a larger equilibrium size group—perhaps at the level of the tribe, where the second level social forces Richerson and Boyd (1999) identified would have come into play. Finally, agriculture is associated with much larger groups; indeed, we have not yet observed the equilibrium size group associated with agriculture and modern technology. Here, both the command and control forces Richerson and Boyd (1999) discussed and market forces are responsible for social coordination.

Note that the key to the evolution of intelligence was competition with other humans, not with other species or with the environment, although intelligence is also useful for coping with the environment. This is shown by the fact that many predecessor hominid species, such as *Homo erectus,* were quite successful and lived (as a species) for long periods of time and in many different physical environments. As Miller (2000, 251) indicates: "Our ancestors were the most powerful omnivores in Africa. . . . But once our ancestors evolved the ability to throw stones, to wave torches around, to attack in groups, and to run for long distances under the midday sun, they were probably the most terrifying animals in Africa. It is a wonder they bothered to evolve more intelligence at all."

At some time, one group of this species evolved into anatomically modern humans, probably with several intermediate species; *Homo erectus* probably did not change as an entire species into modern humans. Increasing evidence shows that this event occurred in Africa about 200,000 years ago and that migration from Africa of modern humans began about 100,000 years ago (Hodges 2000). (For a contrary view, which is however difficult to sustain, see Wolpoff and Gaspari 1997.) Though our intelligence enables us to use tools and other nonhuman elements, this was not the original source of evolutionary pressure.[4] Of course, as intergroup competition became more important, ability to build and use weapons would also have become more important and may have played a role in selection as well. Moreover, increased intelligence is associated with better use of the environment (ecological intelligence) and greater ability to obtain food.

Kaplan et al. (2000) argue that hominid consumption of high quality foods which require more learning and more effort was the major input into intelligence and other aspects of human life history, including long dependency periods and long lifespans. They contrast this with the social intelligence argument made here. But the two are not inconsistent. Social factors would have created a value or demand for increased intelligence; as diets improved, the improvement would have created a supply of resources to satisfy this demand. In this view, both factors were relevant, and there was no need to choose one or the other.

Competition would likely have been both within-group competition and between-group competition. Alexander (1987) argues that the two would have reinforced each other: as between-group competition caused groups to become larger over time, then within-group selection pressure would have increased. This within-group competition would have also lead to increased intelligence. For example, competition for dominance among males and the corresponding increased procreative success would have placed a greater premium on intelligence as group size increased. Moreover, Alexander (1987) and Crawford (1998) argue that if the major environmental force leading to human adaptation was competition from other humans, the current evolutionarily relevant environment may not be very different from the EEA in many biologically relevant respects; both forms of competition still exist. Rubin and Somanathan (1998) have argued that these sources of intelligence are still relevant for understanding human behavior. In a recent paper, Bingham (1999) argues that the major force leading to human uniqueness was the ability of groups of human ancestors to punish cheaters within the group using weapons such as clubs and thrown projectiles. He is able to explain much of human prehistory using this model. The emphasis on weapons and the scope of weapons is quite useful. Miller (2000) argues that sexual selection would also have been a positive feedback force and could have been important in generating human intelligence. His argument is discussed at various points in this book.

The conflict between groups would have been quite violent, and often genocidal. A common goal was the killing of all males in nearby locations. Often the fighting would have been in the form of raids: a group of males from one band would ambush and kill one male in another band and continue this process until all males in the victimized group would have been killed. Again, this is the form of combat among chimpanzees (where about 30 percent of males die violently), but it is also the form that was common among bands of hunter-gatherers.

Women would then generally be incorporated into the victorious band. This pattern was common at least until biblical times (Num. 31):

> (7) And they warred against the Midianites, as the LORD commanded Moses; and they slew all the males.
> (8) And they slew the kings of Midian, beside the rest of them that were slain; [namely], Evi, and Rekem, and Zur, and Hur, and Reba, five kings of Midian: Balaam also the son of Beor they slew with the sword.
> (9) And the children of Israel took [all] the women of Midian captives, and their little ones, and took the spoil of all their cattle, and all their flocks, and all their goods.

Indeed, when considering the recent events in the wars in the former Yugoslavia, the pattern seems to exist even today.

Young Males

Fighters are generally young males. Young males have several characteristics that make them good soldiers, in addition to strength. They are easily persuaded or indoctrinated by official statements or propaganda. They also form groups easily—whether military platoons, sports teams, or gangs. Moreover, in the EEA, combat would have been a rational strategy for young males. If they joined in combat and survived, they would have access to females—either the females of the defeated enemy or from their own group, because the act of victory would have been a sign of fitness (Miller 2000) or because of increased wealth, captured from the enemy. Young males who engaged in combat got reputations for willingness and ability to fight that tended to deter others from challenging them in the future (Buss 1999). If they refused to fight, they would often have been ostracized and would not have left progeny. And they would have had reputations for cowardice, which may have lead others to challenge them or appropriate their mates.

War can pay for individuals in other ways. If potential fighters do not know who will survive but if the expected gains from victory are positive, war will be a rational strategy (Tooby and Cosmides, in press). That is, if a potential combat will kill 20 percent of the aggressors but will provide each survivor with two women captured from the losing side, war can be a rational genetic strategy, and we would have been selected to engage in combat in those circumstances. (Since this strategy has a positive expected payoff, it does not require risk-seeking behavior, which is usually defined as accepting gambles with a negative expected payoff.) Sexual access to the females of the defeated enemy is a standard benefit of victory in combat, both in the past and often today (Ghiglieri 1999; Low 2000). Thus, there was an evolutionary basis for young males being willing to join in fighting and that behavior persists. War in the EEA could be a rational strategy, both for individual warriors and for the band or group. Pressure from this sort of conflict provided selection pressure for increased intelligence. No doubt war was and perhaps remains an important force of selection among humans. (The situation for females was quite different, as discussed in chapter 7.)

One implication of this analysis is that human tastes for defense, and sometimes for offense, are natural. Rational human groups will want to spend some resources on defense against attack by other human groups. Pacifism is not a belief that would have been selected for in the EEA. Moreover, given that offense and attack is sometimes a rational strategy for some groups or members of groups, pacifism is not likely to be a viable strategy even today. On the other hand, because humans do respond to relative prices, such combat is not in any sense inevitable. Indeed, I argue in the next section that ethnic conflict does not pay for most individuals or groups in today's world. As people come to learn this, we may expect reduced amounts of such conflict.

Conflict between groups in the EEA is consistent with the argument in the previous chapter that the world was approximately zero sum. In a zero-sum world, the only way to acquire additional resources is to take them from others. If our ancestors lived and evolved in such a world, those who were more successful at such aggression would have been more likely to become our ancestors. If we evolved in such a world, we might have tendencies to believe that intergroup predation is a useful strategy. Since the world is no longer zero sum, this evolved intuition is now counterproductive, and many are shedding it. I discuss this point in the next section.

Ethnic Conflicts

As I write this, ongoing ethnic conflicts can be found in at least: Ireland, the Middle East, Kashmir, Burundi, and the former Yugoslavia. When you read this (whenever that may be), many of these conflicts will still be active, and others may be as well. Ethnic conflict seems to be a perennial part of the human condition. Relatively small-scale ethnic conflict is the major form of conflict today and is likely to continue to be so. Because members of the same ethnic group are more closely related genetically than are non-members, it may be that ethnic conflict is an evolved human behavior. That is, it may be that by engaging in conflict that is perceived to benefit one's ethnic group, an individual is also providing benefits to copies of his genes that reside in members of the group. If so, this would be an example of kin selection. Even if most members of the ethnic group are only distant relatives, there may be so many of them that the net effect on one's genes of some action that benefits other members of the ethnic group would be beneficial in terms of inclusive fitness. Then ethnic conflict would be a remnant of conflicts in the EEA that had a positive payoff and would be consistent with the argument that struggles and conflicts between groups have been an important part of human evolution. Of course, in the EEA, bands would have been much smaller than the groups that engage in this behavior today. In this sense, such conflict may be the result of a genetic preference that was appropriate in the EEA but is now counterproductive, just as eating excess sugar for a fat person who wants to live longer is an error.

But even if this ethnic conflict was rational and provided benefits in the EEA, it does not mean that ethnic conflict actually provides these benefits today. Rather, the taste for ethnic conflict, which might have been useful in the EEA, may now be counterproductive and may not provide net benefits. This is true whether the benefits are viewed as economic or as fitness benefits; under either interpretation, ethnic conflicts in today's world are harmful and counterproductive to the participants as well as to others. Moreover, they are not inevitable: even if humans have a taste or preference for such conflict, they can be taught that this taste is counterproductive. That is, they can be shown that the relative price of engaging in this preference is too high and that the conflict will provide no net benefits.

The level of ethnic conflict can clearly be reduced. The United States is an example. This is a multiethnic society and, while there is some residual ethnic conflict and some small groups of racists (who sometimes undertake hateful and harmful acts, and who receive a large amount of attention in the media), the level of ethnic conflict is quite small here. The level of such conflict has been greatly been reduced in a remarkably short (perhaps fifty year) time period, as dated from the Supreme Court decision outlawing school segregation. This reduction in conflict has also been responsible for the decreasing racism in immigration policies. Clearly evolution was not involved in this change; learning and perhaps changes in culture and the resulting relative social prices were the driving forces.

Furthermore, ethnic groups in the United States are much more diverse (much less genetically related) than groups in those countries where there is significant ethnic conflict. In Yugoslavia, for example, Serbs, Croats, and Muslims are all Slavs, and the differences between them are primarily religious, not ethnic or genetic; Cavalli-Sforza et al. (1994) classify Yugoslovian as one genetic group. In the Middle East, Jews and Arabs are all Semites; Cavalli-Sforza et al. (1994) point out that the Israelites (ancestors of modern Jews) were originally one of many Semitic-speaking tribes. They also say that residents of Bangladesh are of the same group as some Indians and Pakistanis and are more closely related to some Indians than are other Indians; the difference between these groups is again religion. On the other hand, the United States has large groups of people of African descent, people of Germanic extraction (from Britain and Scandinavia as well as Germany), Mediterranean peoples (Italians, Greeks), Semites (Jews and Arabs), Asians (Indians, Chinese, Koreans), Slavs, and members of many other ethnic groups as well. This diversity is in part due to the greatly increased mobility of contemporary humans, as stressed by Goetze (1998). Our current immigration policy is increasing the level of such diversity. If ethnic conflict were aimed at improving one's genes chances relative to other genes, the United States would be a much more fertile ground for such conflict than places where it actually occurs because there are many more unrelated individuals with whom to engage in conflict.

This is an example of a genetic preference responding to prices, or of facultative behavior. In the United States, ignoring ethnic differences is beneficial, and people respond to these benefits. I argue later in this chapter that ignoring the ethnicity of people is almost always beneficial, and the puzzle is that others have not realized these gains.

Today's world differs in crucial respects from the EEA. As discussed in chapter 1, the EEA was largely a zero-sum world. Our ancestors hunted and gathered, and the amount of food was relatively fixed. If one group killed an elephant or mammoth, that elephant was not available for another group to kill; indeed, its offspring were not available either, so more was lost than just the particular animal killed. If one tribe gathered food in a certain area, other tribes had to go elsewhere. Humans evolved in this environment, and it would have paid for them to evolve preferences against neighbors and for kin and members of the same tribe or clan. Once these preferences evolved (and they probably came from our ape-like ancestors: chimpanzees engage in something very like war and genocide), they would have been self-reinforcing: if I attack members of the neighboring tribe on sight, they will learn to attack me first. If I can identify them by dress or accent or tattoos, these should become useful markers for friendship and enmity. Skin color would probably not have been relevant in the EEA. Neighbors with whom one fought would have been relatively close relations (relative to the entire genetic variation of mankind) and of the same color. This is still true today. The combatants in the disputes mentioned earlier (Catholic Irish and Protestant Irish, Jews and Arabs, Indians and Pakistanis, Tutus and Hutsis) are close relatives compared with the entire range of mankind and cannot be differentiated by skin color.

Ethnic conflict is not fitness improving in today's world for at least three reasons. First, for most of the world, land, the subject of ethnic conflict, is not the major constraint on fitness or on wealth. Second, in many cases, the rest of the world punishes groups for engaging in ethnic combat, so the net result is not an increase in fitness or in wealth. Third, the modern world is not zero sum. Rather, there are possible gains from trade in many dimensions. Ethnic conflict reduces the possibility of these gains and thus harms the combatants. I now discuss each point.

Land

In today's world, the issues relevant in ethnic conflict are not related to fitness for most people. In the EEA, conflicts were often over land or territory, and land was a major input into survival and thus into reproduction. Today, land may be valuable, but it is only one valuable asset among many. Possession of land is only tenuously related to the ability to provide food. A country such as Japan with little land per capita can easily feed its people by trading other goods (cars, televisions) for food. (Such trade is indirect: Japan sells cars on the world market and uses the money thus obtained to buy food.) A country such as Russia with huge amounts of arable land cannot feed itself because its economic and legal systems are so disorganized that the country cannot function effectively.

Land may still be valuable, and fighting over land on occasion may be worthwhile. But land is no different than any other resource, and in many cases the value of land can be destroyed by the very conflict. Some of the value of land comes from minerals, which can be captured through conquest. But much comes from the organization of production. The value of real estate in New York City, London, Tokyo, and Hong Kong does not come from the minerals under the land or from the crops that can be grown. Rather, they are valuable because of the buildings and, more important, because of the human capital of the workers in the buildings that is agglomerated in the cities. An invading army conquering one of these cities would not be able to appropriate the wealth that human capital creates by taking over the land. Rather, to benefit, a conqueror would be forced to allow the existing population to continue working. Any benefit to the conqueror would come from allowing production to continue and taxing it, as in the case of China and Hong Kong. This strategy is very different from massacring the men and raping and kidnapping the women, the standard form of ethnic conflict in the EEA.

Moreover, for many people (perhaps most people), biological fitness is not itself a goal. We do not have the maximum number of children that we could biologically support. In many wealthy countries in Europe and in Japan, population growth is actually below replacement levels, and this is not due to wealth constraints (Hrdy 1999; M. Singer 1999). In China, draconian population policies are aimed at reducing the number of people in the country. Indeed, since the Chinese have a preference for sons, population policies will lead to further reductions as there will be insufficient females for the number of males in future generations. As other countries become wealthier, we may expect similar patterns. The constraint on the number of children comes from parental desires or social policy, not any economic limit. Policies that would have led to increased fitness in the EEA will not have that effect today. Thus, again, land is valued for its economic value, not for its ability to grow crops to increase fitness. Hirshleiefer (1998) also points out that modern war does not serve direct fitness purposes.

Punishment

In many cases, the world punishes those who engage in ethnic conflict. This punishment is not certain and does not always occur. But it is probably common enough so that it means that ethnic persecution or conflict does not pay, particularly given the relatively small potential gains from this conflict in today's world. For one example, a worldwide boycott of South Africa ultimately caused the ethnically based apartheid regime to collapse. For another, the Serbian policy in Kosovo was quite costly to the Serbs as a result of U.S.-led NATO bombing. Indeed, Serbian policies have actually led to a decline in fertility. From 1990 to 1998 the number of babies born in central Serbia declined by 15 percent. The population growth rate has fallen in the 1990s from a 1.2 percent annual growth rate to a 1.8 percent decline (Harden 1999). Neither Germany nor Japan, both of which practiced horrid ethnic policies in World War II, gained from these wars. Both countries are prosperous today, but this prosperity comes from trade and exchange with ethnically heterogeneous neighbors, not from exterminating them. In Germany, many productive workers from Turkey, who are not very closely related to the Germans, make their home there, and the country is beginning to seek additional immigrants (Zachary and Rohwedder 2001). El Qaeda and their supporters in the Taliban in Afghanistan did not benefit either in economic or fitness terms from the religious conflict against the

United States. Of course, in some circumstances, leaders of countries may still benefit from ethnic conflict, but the policies are probably not in the genetic or economic interest of the citizens of the aggressor countries. War crimes trials for leaders of defeated aggressors, however, may serve as a useful deterrent.

Gains from Trade

The third reason ethnic conflict does not pay is because this conflict eliminates potential gains from trade. These gains come from two sources. First, if members of ethnic groups have similar skills and abilities, simply expanding the size of the market can be beneficial. Second, if members of different ethnic groups have different skills and abilities, there are gains from trade.

Consider first the size of the market. As Adam Smith long ago pointed out, "the division of labor is limited by the extent of the market" (1789/ 1994, 19). This means that as the number of people in a market expands, each worker can specialize in a narrower area of production. As this happens, productivity (real wealth) increases. More people in a market also create the possibility for more specialized consumption opportunities. It is also true that increased population leads to a greater possibility for technological advance, as shown by Julian Simon (1996), Michael Kremer (1993), and Charles Jones (2001). Of course, additional people can be from the same ethnic group, but if population growth has slowed or become negative, as seems to be happening in many wealthy countries, the only way to realize these gains is to allow members of different ethnic groups to join the society. For goods, there is the possibility of import from other countries; that is, the market for goods may be worldwide, so that a larger population in a country is not needed to produce goods. But services are produced domestically, and the share of services in world GNP increases with wealth. Thus, the possibility of international trade does not contradict the point made here.

The second type of gain from trade occurs if the members of different ethnic groups have different skills and abilities, for whatever reason these differences may exist. In that case, there are possibilities of members of each group specializing in different activities and then engaging in mutually beneficial trade. Of course, in this circumstance, those particular members of the dominant group who compete most closely with members of the smaller group would have incentives to reduce this trade, perhaps by generating ethnic conflict, but members of the society overall will benefit from the increased trade. For example, in the United States, immigration of trained scientists, computer programmers, and mathematicians from India, China, and the former Soviet Union has greatly increased our pool of skilled work-ers and is a significant part of the current economic and technological boom. One benefit of reduced ethnic conflict is the ability to profit from increased immigration of ethnically diverse individuals. For the United States, an additional benefit is that immigrants will be able to maintain our Social Security system as demographic changes of the existing population put pressure on this system (Storesletten 2000). Nonetheless, some Ameri-can mathematicians have protested about the competition from Russian mathematicians.

To make this point more carefully, I will consider Jews as an ethnic group. Jews have been a minority ethnic group in many societies and have been persecuted in many of these societies. Moreover, a recent series of books by an evolutionary minded scholar, Kevin MacDonald (1994; 1998a; 1998b), has analyzed anti-Semitism from an evolutionary perspective, but he has missed the key point about increased gains from trade. Therefore, exami-nation of this ethnic group will provide a useful demonstration of the gen-eral point. The general assumption is that Jews will live in a society as a minority, where the ethnic majority is from another background.

MacDonald (1998a, vii) argues that anti-Semitism is caused ultimately by "resource competition and conflicts of interest" between Jews and gentiles living in the same society. The key point that is largely missed or ignored is that this resource competition will generally provide economic benefits to gentile society overall, even if it does harm some segments of gentile soci-ety. MacDonald sometimes indicates that there are benefits from trade with Jews to some gentiles, but only in passing. For example, "anti-Semitism is ex-pected to be . . . least common among gentiles who are actually benefiting from the Jews, such aristocratic gentiles who often profited from coopera-tion with them" (9). That Jewish merchants, doctors, teachers, and money-lenders also provide benefits to their customers, patients, students, and debtors is not mentioned.

Though I do not accept it, I will, for the sake of argument, grant Mac-Donald's thesis, that Jews are a separate society within gentile society. (This may have been more appropriate in medieval Poland than in contempo-rary America.) However, in his analysis, these two societies engage in eco-nomic exchange. An appropriate model to use in analyzing this situation is the model of trade between separate countries, the model of interna-tional trade. The results of using this model are unequivocal: members of both societies, Jew and gentile, gain economically from free exchange. (This use of the model of international trade to discuss ethnic groups in a society is not original: Gary Becker 1971, uses this model in discussing black-white discrimination and shows that both groups benefit from exchange or lose from segregation.) Those groups or individuals in direct competition with Jews might lose from this competition, just as some American firms and workers lose from international competition with other countries. But the broader point is that the gains to those who gain from exchange (the customers of the Jewish merchants or the clients of the Jewish moneylenders) gain more than the losers (non-Jewish merchants or moneylenders) lose, so overall there are net gains to gentile society from free trade. This is a standard argument used to analyze trade: American consumers gain more from international trade than competing American producers lose.

These gains exist even if Jews could succeed in colluding and acting like a cartel, or a monopoly, which is itself highly unlikely. Normally, economists believe that large groups of sellers, such as all the Jewish merchants in Poland, or even in Warsaw, cannot successfully collude because of free-rider problems of the same sort that some argue may make group selection impossible.[5] But even in this worst case scenario, gentiles benefit from trade with Jews. For retail trade and most other businesses, the limits of the ability of a Jewish monopoly to exploit gentile customers are given by the possibility of other gentiles entering to compete. If Jews are more efficient than gentiles, Jewish monopolists will make profits. Nonetheless, gentile customers will still benefit from trading because the Jewish merchants still sell goods for lower prices than would gentile merchants, else gentiles will enter the market. Jewish merchants might have been more efficient, perhaps because of effective links between countries. However, even for a monopolist, if costs are lower, the profit maximizing price will be lower, so efficient traders would have benefited their customers as well as themselves. Greif (1993) discusses such links for the eleventh-century Jewish Maghribi traders. He indicates that these traders lost their markets when the Italian merchants became more efficient. This notion does not support MacDonald's argument that Jewish traders were monopolists.

Take the stronger case. Assume that gentiles are forbidden from engaging in moneylending, perhaps by religious prohibitions, and, therefore, Jewish moneylenders have no competition. (This position is actually false: Barzel 1992, shows that the Jewish moneylenders in England lost their market to the Italians in the thirteenth century.) Assume also that all Jewish moneylenders collude and charge the monopoly price. Nonetheless, even under these extreme and highly unlikely conditions, gentile borrowers still benefit relative to the situation with no moneylenders. That is, dealing with a monopolist is better than having no opportunity to trade at all. The terms of trade for dealing with a monopolist are worse than the terms of trade in dealing with a competitive firm, but some voluntary trade is better than none.

If my argument is correct, anti-Semitic societies should have been harmed by their unwillingness to deal with Jews. In discussing the Inquisition, MacDonald (1998a, 15) indicates that "The Inquisition had a very chilling effect on intellectual endeavor in Spain for centuries." With respect to Roman Christian anti-Semitism, another of his case studies, he provides no discussion of the effects. However, the period following the imposition of restrictions on Jews MacDonald discusses is sometimes called the Dark Ages, indicating that it may be characterized by relatively little intellectual advance. With respect to the Nazis, it is interesting to note that the American nuclear bomb was produced in part with inputs from many Jewish scientists who left Germany and other European countries in fear of the Nazis. Without Nazi anti-Semitism, some of these scientists might have worked on the German war effort. Thus, it is at least plausible that anti-Semitism has had the negative effects on the gentile community that economic theory would suggest.

Jews would also be harmed by separatism, and the harm to Jews would be greater than the harm to gentiles. This is because there are fewer Jews and so a smaller internal Jewish market and less possibility of specialization within the Jewish community than within the larger gentile community. Thus, the loss of trade opportunities with gentiles will harm the Jewish community more than the loss of trade opportunities with Jews will harm gentiles. Self-sufficiency is more expensive for a smaller group. But it is costly for both groups.

It might appear that economic arguments are irrelevant in an analysis of fitness. Fitness is often defined in relative terms, and economics deals with absolute incomes. This is a serious issue. But it is not relevant here. Most populations at issue, Jewish and gentile, were growing over the relevant time period. Thus, if trade with Jews would have benefited gentiles, then gentile population could have grown faster than otherwise. Moreover, competition between gentile populations existed during this period. Any gentile population that established economic relations with Jews would have been expected to benefit relative to other non-Jewish populations.

Conflict between neighbors was an important fact during human evolution; as discussed earlier, some think that this conflict was one of the major inputs into human evolution. Nevertheless, exchange is now also important and there are gains from exchange. It is a common error to view exchange in competitive terms; discussion of trade wars is a staple of the popular press and of politics. This is an example of the sort of zero-sum thinking discussed earlier. The implicit notion is that there is only a certain amount of money to spend, and if it is spent with foreigners, it is not spent with domestic firms. This notion is wrong. Its seductive power is such, however, that even a scientist such as MacDonald has erred in understanding the issue.

Affirmative Action

Shaw and Wong (1989) argue that patriotism with respect to a country can conflict with group or ethnic identity if these markers are more comparable within ethnic groups than within a country. They also suggest that in the United States, identification with the nation is strong; since American citizens are descended from immigrants, there is little ethnic group identification to compete with patriotism. On the other hand, they suggest that in the (then) Soviet Union, patriotism was generally weak because there was strong ethnic group competition. Subsequent events seem to have proved them correct, at least with respect to the Soviet Union and some of its former satellites, which have partially disintegrated largely on ethnic lines and seem to be continuing to disintegrate. Because Shaw and Wong (1989) were concerned with war, however, they seem to have missed or ignored other nongovernmental potentials for group loyalty and identification. Hirshleifer (1998) also discusses the rule of group identification in war.

Shaw and Wong's (1989) analysis of the United States suggests that the country benefits from unconcentrated ethnic identification. This is correct, as discussed also in Rubin and Somanathan (1998). That is, the United States is a stronger and richer country because there is relatively little internal ethnic strife or conflict. Part of the reason for this is because of the large number of ethnic groups here. Most old countries, including European countries, tend to have at least a majority core of individuals from the same ethnic group (Germans, French, Italians). Some countries have two or more ethnic groups (Switzerland, Belgium, the former Yugoslavia), but membership is well defined and geography often separates them. These different ethnic groups can sometimes lead to major conflicts (again the former Yugoslavia.) The United States has the advantage of being composed of many groups, so there is no majority, and whatever geographic isolation exists is largely on a local level. That is, ethnic enclaves may exist within cities but not across states. This means that the United States is very difficult soil for ethnic conflict, and this is a real benefit to our society. The only major internal conflict in the United States, the Civil War, was fought between members of essentially the same ethnic group, in part for the benefit of members of another group.

There is, however, some danger that current policies are acting to strengthen ethnic identification, with potentially harmful results. In the United States today, affirmative action is a massive program whose purpose is to change the outcome of decisions in school admissions, hiring, and other social processes. The goal is to achieve a more race-balanced outcome. But the method involves treating individuals as members of ethnic groups rather than as individuals. This is a very dangerous policy and has commonly led to ethnic polarization wherever it has been adopted (Sowell 1990). While the group identification mechanism is flexible, it is also powerful, and if membership in an ethnic group becomes important for significant purposes, this membership can easily become the basis for strong group identification.

One of the great accomplishments of the United States is the reduction in the importance of race and ethnic background in decision making, both private and public. This has not been easy. At various times, ethnic groups in the United States have been the target of discrimination. Laws and policies have discriminated against the Irish, Jews, Asians, Catholics, Hispanics, blacks, and many other groups, and of course at one time blacks were slaves. Only in 1954, with the Supreme Court decision outlawing school segregation in *Brown v. Board of Education,* did we begin a strong national effort to eliminate official race-based discrimination (see Sowell 1983). In a remarkably short time, we have made a major successful effort at accomplishing this goal.

The history of human evolution tells us that group identity and conflict are a common part of human behavior, a behavior that is easy for humans to learn. Eliminating many aspects of this behavior has been a major accomplishment of policy. But the victory has been fragile. Given the evolved nature of human preferences, it is all too easy to establish a system of group identity that is harmful. Affirmative action programs base important life decisions with substantial economic and fitness consequences, such as college admissions and hiring, on race. This policy creates the possibility of renewing grouping and self-identification based largely on race. Beneficiaries identify as members of an ethnic group because they are treated as group members and benefit by being members. Members of the majority feel that they are being discriminated against, and so have incentives to act as members of a group to defend themselves.

These decisions deal with important matters that are analogous to issues that would have had fitness implications in the EEA, such as level of incomes or control of resources. Therefore, we can expect those who feel harmed by the decisions to react strongly. This reaction is not due merely to losing in economic or political competition. Rather, it is the result of a belief that the competition was ethnically based and, therefore, unfair. Even those who do not directly lose from these policies may therefore resent them. The real danger from affirmative action is that it can unleash the very ethnic identification and ethnic-based policies that it is aimed at correcting. This could be a very harmful policy to adopt. The importance of ethnic conflict in human history and prehistory tells us that it is a dangerous force to unleash, even if the reasons are of the best sort. Indeed, workplace racist incidents seem to be increasing; this increase may be a result of affirmative action

(Siwolop 2000). This same criticism cannot be made of affirmative action policies based on gender, although this should not be taken as an endorsement of such policies.

Summary

Humans universally distinguish between in-group members and others. Though this distinction is made almost automatically, the nature of groups is very flexible. Humans can identify with almost any group, no matter how arbitrarily defined. Such group identification is based on tribal mechanisms in the EEA, where there was substantial conflict between bands or groups. This may be the basis for the ethnic conflict that we observe around the world. However, the example of the United States and other multiethnic societies indicates that such conflict is not inevitable, but rather can be reduced. This example would be useful for those societies that are currently experiencing such conflict, particularly since such conflict in today's world is not productive in any sense—either in terms of increased fitness or in terms of increased wealth. For example, societies that engaged in anti-Semitic behavior suffered from this behavior, as theory would predict. The United States has the advantage of being composed of citizens from many diverse ethnic groups, and ethnic conflict is relatively limited here. But policies in the United States and other advanced countries that stress race are counterproductive and might well decrease or eliminate the remarkable gains the United States has made in reducing ethnic competition.

CHAPTER 3

Altruism, Cooperation, and Sharing

In the previous chapter, I discussed conflict. Here, I turn to better elements of human preferences—altruism and sharing. Just as there are inherited preferences for conflict, so we see that there are also preferences for cooperation and generosity.

Altruism and Group Selection

The basic problem of sociobiology is the problem of altruism, or cooperation. The problem is this: Assume two groups of animals in the same species. In Group A, animals cooperate and assist each other. In Group B, they are selfish and do not cooperate or behave altruistically. Then Group A will be more successful or more fit than Group B. The population of individuals from the group with cooperative tendencies will grow faster than the population of the noncooperative group, and over time the entire population of the species will become composed of descendants of the altruists in Group A. In this reading, altruism is not surprising, and we should expect all social animals to be largely altruistic and cooperative. This is group selection.

BUT: Consider one animal in Group A, the group of cooperators. If one animal does not cooperate (cheats, shirks), it will be even more successful than others because it can free ride on the cooperative behavior of the others and so obtain even more benefits than the cooperators. That is, it will obtain the benefits provided by the cooperators but will save the costs of cooperating itself. Then the cheater will do better than the cooperators in the group of cooperators, and thus will be more fit—will leave more noncooperating offspring. Ultimately, cooperators should be eliminated from the group and from the population, and only noncooperators should be left. This is individual selection, and it is generally thought that individual selection is stronger than group selection. That is, both individual and group selection exist, and selection occurs at both levels—it is multilevel.

But we do observe some cooperators and much social cooperation; this is the puzzle of altruism. The altruism debate in biology goes back to at least the 1960s. In 1962, Vero Copner Wynne-Edwards published a famous book in which he argued that animals behaved altruistically and undertook many activities for the good of the species—for example, by limiting their rate of population increase to avoid overpopulation. The story was one of group selection. That is, groups competed with each other and that group which was more cooperative would succeed and take over the population. The stories were very agreeable and pleasing, and the notion of altruistic and kindly animals behaving nicely toward each other appealed to many. The only problem was that the analysis did not make any sense, for reasons mentioned earlier: any one animal would have gained by cheating and ignoring the altruistic consensus. George C. Williams, in a very important book published in 1966, made this point strongly and emphasized the role of the individual in natural selection. Richard Dawkins, in a famous book published in 1976, argued that the unit of selection was not even the individual, but rather the selfish gene. As a result, any argument that smacks of group selection has been taboo in biology. The only acceptable theoretical stance is one of individual selection, or sometimes gene selection. The puzzle is to explain altruistic behavior without resorting to group selection.

The problem has been modeled in game theory as the prisoner's dilemma, the most famous game in the theory and a fundamental form of analysis for the social and biological sciences. (For a readable analysis, see Ridley 1996, chapter 3.) The original story of the game was this: The police have arrested two criminals for some minor crime—say, possession of drugs. The sentence is three months. They suspect that the two also robbed a bank. They then question each prisoner separately and offer this deal: If you confess to the bank robbery and your partner does not, then you will get off and your partner gets eight years. If your partner confesses and you don't, then he will get off and you get the eight years. If you both confess, then you each get five years. If neither confesses, you both get the three months for possession.

Consider the strategy from the perspective of one prisoner. (Since the payoffs are symmetrical, each is faced with the same choices.)

> If he doesn't confess and I do, then I get off and he gets eight years. If he doesn't confess and I don't confess either, then we each get three months. Therefore, if he does not confess, my best strategy is to confess since that reduces my sentence from three months to zero. On the other hand, if he does confess and I don't, I get the eight years. If he does confess and so do I, I get five years. Therefore, if he does confess, I am better off confessing as well since I reduce my sentence from eight to five years by confessing. In other words, no matter what my partner does, I am better off confessing.

Since the game is symmetric, both players confess and both get five years instead of the three months they would get if neither confessed. Rational behavior on the part of each party leads to the unfortunate (for the prisoners) outcome of both getting larger sentences. (To see this game in action in its original context, watch the television show *NYPD Blue* almost any week.)

The structure of the game is very general. It depicts a situation in which cooperation is better than defection (cheating) for both parties, but for each party, defection is the rational strategy. The same story can be told about contributing to a public good such as defense: If we both contribute, then we are both better off. But no matter what your strategy, my best strategy is to not contribute and be a free rider, or shirker, or slacker. This model is very general because it can be applied to so many situations. Within a group selection model, the story is that the entire group would do better if members cooperated, but each member would do better by not cooperating. The logic of the prisoner's dilemma governs, and there is no cooperation—there is no group selection.

Theorists have identified several mechanisms that may ameliorate the prisoner's dilemma problem and lead to some cooperation among humans and other animals. I now discuss these mechanisms.

Mechanisms for Social Behavior in Humans

This question of how to explain social behavior and cooperation is particularly difficult for humans, because we are very highly social and live in much larger groups than any other animal. Biologists have analyzed two major mechanisms for social cooperation, or for ways around the prisoner's dilemma. First is kin-based altruism, discussed in chapter 1. However, for large groups, kin-based altruism is clearly insufficient because relatedness falls off quickly. Biologists have invented another mechanism to explain some forms of altruism, reciprocity. In a noted paper, Robert Trivers (1971) discovered contract in a biological setting and provided a biological basis for exchange. Essentially, reciprocity is a mechanism under which one party provides benefits today in return for benefits returned tomorrow. For reciprocity (or reciprocal altruism) to work, certain requirements must be met. For one thing, animals have to be long lived, so there is a reasonable expectation of receiving a return on an investment. For another, they have to be sufficiently intelligent both to identify specific other animals and to recall benefits received from such animals. Humans and some other advanced species (including apparently vampire bats) meet these requirements. Trivers also identified some mechanisms for enforcing reciprocity, including psychological mechanisms such as anger at being cheated and other forms of moralistic aggression. Nonetheless, although reciprocity was clearly part of the human evolutionary past, and we do behave as Trivers has suggested, Boyd and Richerson (1989) have shown that this mechanism will work only for small groups.

Other mechanisms that can generate some altruism, or reduce the amount of free riding, have been identified. One point to note is that the actual structure of interactions in real populations is not a one-shot prisoner's dilemma. Rather, long-lived animals (including humans) will interact with the same individuals many times over the course of a lifetime, particularly in small groups such as predominated during the EEA. This situation leads to a difficult mathematical problem, the problem of the repeated game. (For an accessible analysis in a biological context, see Skyrms 1996 or Gintis (2000a.) One theorem in game theory (called the folk theorem) is that essentially anything can happen in such games. That is, one equilibrium is cooperation; other equilibria include complete noncooperation. There are an infinite number of possible equilibria, so the problem—which we address in more carefully in chapter 6—is selection of the appropriate equilibrium. The mathematics tells us that cooperation is possible, but it does not provide mechanisms for getting there.

In one famous set of experiments, Axelrod (1984) showed that a strategy called tit for tat could be successful against many opposing strategies in a repeated prisoner's dilemma setting. In this strategy, I play this period whatever strategy you played last period. If you cheated last period, then I cheat this period. If you cooperated, then I also cooperate now. Thus, we can get a sequence started in which we both cooperate each period. Humans may be selected to play such strategies or variants of them because our predecessors who did were more likely to become our ancestors.

Another way of generating cooperation is through an asymmetric interaction. An example is Maynard Smith's (1982) bourgeois strategy, discussed in chapter 1. In a game called hawk-dove, hawks always fight and doves always give in. Then in a population of doves, hawks will do well, because they will almost always win without a fight as the doves concede. But as there are more hawks, they begin to do poorly because they fight with each other. A third strategy, bourgeois, is that in which the first player to arrive at the resource wins. The strategy is: play hawk if you are first and dove if you are second. This strategy beats both hawk and dove. This strategy treats the players asymmetrically—the first to arrive is the winner. Some people see this as the basis for property rights.

Two other possible mechanisms rely on the intelligence of humans. One very important fact is that humans can choose with whom to interact. If someone is known as a cooperator, others will want to interact with this person. But the cooperator will himself want to interact with other cooperators. In this situation, there can be assortment by amount of cooperativeness. Then groups formed by more cooperative players will do better than groups with less cooperative members, and members of the cooperative group will have more offspring. As a result, the degree of cooperativeness can grow in the population over time. This assortment by cooperativeness has another interesting feature. Everyone (cooperator or cheater) would prefer to deal with a cooperator. Therefore, individuals will have an incentive to appear to be cooperators even if they are not. Those who are not cooperators will try to fool others into believing that they are by mimicking cooperators. But this leads to an arms race: as people mimic cooperative behavior, people will also generate detection devices to determine who is cheating and who is really a cooperator. The result is that humans are both very sophisticated cheaters and very sophisticated detectors of cheating (Frank 1988). Indeed, we seem even to have evolved specific mental mechanisms for detecting cheating (Cosmides and Tooby 1992). Even more: the best way to convince someone else that you are a cooperator is to first convince yourself. As a result, humans turn out to be very good at self-deception (Trivers 1971). Some implications of self-deception are discussed in chapter 7.

The second mechanism that relies on intelligence is the possibility of punishment. Humans can punish those who violate rules, including free riders. De Waal (1996) has shown that chimpanzees also punish cheaters. Experimental evidence indicates that humans are quite willing to punish cheaters in public goods settings (see, e.g., Fehr and Schmidt 2002). Second-level punishment is even possible among humans, but not to my knowledge among other animals: humans can punish those who refuse to punish those who violate rules. Second-level punishment can generate almost any behavior among humans, whether efficient or inefficient (Hirshleifer and Martinez-Coll 1988; Boyd and Richerson 1992; Axelrod 1997; Gintis 2000b). In the EEA, punishment might have been relatively inexpensive, as it could have taken the form of banishment. Today, some forms of social punishment, such as social ostracism, are inexpensive; others, such as a trial and formal prison time, are rather expensive. Bingham (1999) has built a complete theory of human evolution on the notion of punishing cheaters in this context. He argues that a fundamental human adaptation was the ability to physically punish cheaters using clubs or thrown missiles (e.g., rocks.) In a recent paper, Hirshleifer (1999) discussed these and other pathways for the evolution of cooperation and indicated that biologists err by emphasizing only kinship and reciprocity.

Alexander (1987, 85; see also Krebs 1998, 347) discusses indirect reciprocity, which is a complex multilateral system of reciprocity. Boyd and Richerson (1989) show that this system would be difficult to sustain in a system of networks, each containing several individuals. This analysis, however, would be greatly simplified if what is called indirect reciprocity was viewed as multilateral exchange within a general equilibrium framework and that, as economists have known, everyone can benefit from such exchange systems.

A functioning market economy, of course, coordinates huge amounts of cooperation. Almost every act by a person in such an economy involves cooperation by literally millions of others, from all over the world. While capitalist economic systems are called competitive, in actuality they are the largest cooperative endeavors in the world. Many have erred in confusing this point and focusing on the competitive details of such economies, thus neglecting the cooperative aspect (e.g., Singer 2000). This is clearly an area of research where the biologists would benefit from inputs from economists. This is an additional economic principle in human evolution, in addition to those (Smith 1991) identified.

Richerson and Boyd (1999) suggest that two separate mechanisms cause social cooperation. Smith (1998) has a similar analysis. One is the evolutionarily old cooperative mechanism shared to some extent by all primates. This operates primarily in face-to-face environments and among small sets of individuals and is based on kin selection and on reciprocal altruism, in the sense of Trivers (1971).

The other is unique to humans and involves much larger networks of exchange and cooperation. This mechanism is the basis for the large markets in modern economies. Such mechanisms may need outside enforcement, as through a court system with the possibility of police to back up the court's orders. It is the subject matter of economics and raises an interesting question: How important are evolutionary adaptations for functioning in large, impersonal markets with either endogenous or outside enforcement mechanisms? The answer to this question will be important for determining the influence of evolutionary thinking on the core of economics, political science, and law.

Both mechanisms still operate, but in different contexts. For example, criminal organizations use the former mechanism (Richerson and Boyd 1999), which is useful for limiting the size and power of such organizations. Larger mechanisms include cooperation among tribal-sized entities. Rubin (1994) discusses such mechanisms in the context of Russia and other economies lacking legal institutions (see also Landa 1981, who stresses exchange among kin and members of the same ethnic group). Keeley (1996, 123) indicates that an important source of war in primitive societies is the failure of recipients to effectively reciprocate, which he indicates is the equivalent to default or fraud in a more commercial system. More generally, Keeley discusses the hazy border between trade and extortion in an environment with no clear enforcement devices. In this view, large modern institutions are made up of smaller components consistent with the evolutionary background coupled with enforcement devices, because mechanisms consistent with this large scale have not had enough time to evolve. This is one example of the difference in scale between modern and evolutionary relevant societies, as mentioned in chapter 1.

Group Selection

A major debate in biology and, in particular, in the study of human evolution is the extent of group selection. Since the work of Williams (1966), it has generally been thought that any form of group selection was impossible because free riding within the group eliminates the possibility. More recently, in a series of articles and an important book, Sober and Wilson (1998) have reignited the debate. They argue that in some circumstances, groups containing more altruists would grow faster than groups with fewer altruists, even though the number of altruists within each type of group would decrease over time. If the groups remain isolated, the standard result would obtain, and the number of altruists in each group and thus in the population would ultimately go to zero. However, if the groups periodically split and re-form, the number of altruists in the population could increase. The argument is that we must consider carefully the level of selection. They also argue that this mechanism would have been particularly important in human evolution and present some data indicating that human behavior is consistent with their argument.

None of the mechanisms discussed in the previous section are as powerful at generating altruism as complete group selection.[1] Therefore, if we can determine the types of altruism that group selection favors, we can understand the limits of altruism. If some suggested form of altruism is not compatible with group selection, it is unlikely that humans have evolved a preference for this behavior. Of course, the converse may not be true: if some mechanism is consistent with group selection, this does not prove that this mechanism exists unless we are confident that some form of group selection actually occurred. In what follows, I take no stand on the existence of group selection, which is quite controversial among biologists. Rather, I provide a "what if" argument: if group selection exists, what is the strongest form of altruism that it could support? We may then be confident that our actual evolved preferences are for no more (and perhaps for less) altruism than this.

The other point to note is that most agree that the EEA was an environment of intense group competition, as discussed in chapter 2. Predation by hominids on hominids or by humans on humans was a very important factor in the EEA. Many analysts believe that this selection pressure was responsible for the tremendous increase in intelligence and brain size as our ancestors evolved from an ape-like creature to become humans. Thus, in evolutionary times, group size was important. That is, having more individuals in the group was a positive externality for each member. This is a consequence of the balance of power argument—that the purpose of sociality was defense against predators, primarily other hominids (Alexander 1987). Alexander denies that his argument relies on group selection, but this may not be correct (Wilson 1999).

In some circumstances, seemingly altruistic transfers to some individuals would have been in the group interest, and presumably in those circumstances the utility or well-being of the recipients would have been an element in the utility functions of the donors. The main point is that transfers would have been desired in situations where a low-cost transfer could save the life of the recipient and enable him to continue to participate in the group defense or enable her to reproduce other defenders. Two classes of situations would have been relevant in the EEA and they are still relevant today (because our tastes evolved in the EEA). These are temporary income shortfalls, which could lead to starvation, and illness or injury, which could lead to death. Both are discussed later in this chapter.

Another mechanism, in addition to direct competition, would have favored more efficient preferences. Mobility between groups was often possible in the EEA (as it is today). Groups following more efficient policies would have been richer (had control over more resources), and in many cases, individuals would have voluntarily chosen such groups. This raises the issue of cultural rather than biological selection of preferences. The most useful models of such selection processes allow genes and culture to interact (see Boyd and Richerson 1985). In their model, there is evolutionary interaction between genes and culture. For example, if cultural norms in some group favor some form of altruism, individuals who most easily adapt to this form of altruism will be more successful in that group. Cultural and genetic evolution can reinforce each other. But in this book I focus on the genetic part of the process.

Efficient Altruism

The analysis of group selection leads to the notion of what I call *efficient altruism*. In the Sober and Wilson analysis, groups with altruists grow faster, and this explains the survival of altruism in the population. But only some types of altruism would generate this differential growth rate. Group selection is not an indiscriminate mechanism for generating everything that is called or appears to be altruism, or for indiscriminate generosity. Rather, some behaviors that some have called altruistic might actually lead groups that practiced these behaviors (or that had many individuals practicing these behaviors) to grow more slowly. Natural selection could not have generated preferences for such behavior in humans. I call altruism that does lead to increased growth of groups with more such altruists (or that would have done so in the EEA) efficient altruism.

The general example of such efficient altruism is cooperation in some variant of a prisoner's dilemma game. The prisoner's dilemma is the generic cooperate-don't cooperate game, where cooperation is efficient for the cooperators jointly in that it leads to greater payoffs for both participants, but noncooperation is the best strategy for each player. Then some group in which more people played cooperate would have greater income and wealth than some group where fewer individuals cooperated, and the population in the more cooperative would grow more quickly. Sober and Wilson (1998, chap. 5) show that most or all societies enforce social norms. One important function of such norms is to induce cooperation and reciprocation—that is, to force a cooperative solution to a prisoner's dilemma like game.

Cooperation: Examples

I now discuss some examples of behaviors that are consistent with group selection. I do not want to imply that I believe that group selection has operated in humans; the evidence is still out and I am agnostic on this issue. The behaviors I describe are consistent with group selection, but may be consistent with individual selection as well. The point is really the converse: Preferences for any allegedly altruistic behavior that is not consistent with group selection (and I discuss some later) could not have survived. That is, group selection is sufficient but not necessary for these behaviors.

One example of efficient altruism is be cooperation in the purchase or creation of public goods. In the EEA, many things that we think of as public goods (roads, public buildings) would not have been meaningful. (I discuss the role of inherited altruistic preferences in creating of contemporary public goods below.) But law and order—enforcement of rules—would have been a useful goal. Much of the discussion in Sober and Wilson and others about rule enforcement (punishment for rule violation) and second-level rule enforcement (punishment for not punishing rule violations) may be construed as contributing to a public good. Thus, one form of altruism that may be favored in human populations is providing institutions or mechanisms for rule enforcement. Rules and rule enforcement would lead to larger incomes in groups and to exactly the sort of differential growth that is required in the Sober and Wilson model. Rule enforcement is universal among humans (Brown 1991). I discuss the nature of the rules themselves below.

Cooperation in productive activities is also be an efficient form of altruism. An example is group hunting. Cooperation would be to do one's share even if the activity is dangerous; shirking or noncooperation would be to free ride and not do one's assigned task. Then if all cooperate, the hunt will be more successful—this is exactly a prisoner's dilemma. Groups with hunters who were more cooperative (or, more generally, workers who were more cooperative) would grow faster than others, and, therefore, this form of altruism could survive in the EEA in a Sober and Wilson process. Ridley (1996) indicates that big game was the first public good.

Similarly, if group conflict was an issue in the EEA—and I have indicated elsewhere that many think that such conflict has been an ongoing part of human and even prehuman existence—contributions to defense or offense would also have been a public good. The main form of contribution to this public good would probably have been rewards to individuals, largely young males, who participated in defense or predation, and evidence indicates that such rewards are in fact forthcoming. That is, the public good would have been the benefits provided to warriors to induce them to participate in militaristic activities. Thus, while individuals may have participated in defensive or offensive military activities for private motives, the reason that private motives would have been available is because such participation brought public rewards. Cooperation by young males in offensive or defensive activities is a form of altruism that would have survived a group selection process. These preferences survive as participation in gangs or sports and are an important reason why young males are a major component of military forces. There appears to be a mental module for such cooperation by young males.

Another form of altruism—one more consonant with the normal usage of the term—would have been sharing of food or other resources. Food sharing could again have been a form of efficient altruism. If one person is low on food today he may die. If another shares his food, this will have desirable effects on the group, since it will lead to a larger group size, which is useful for protection. Additionally, food storage would have been difficult. Hunting would have been an occupation with a relatively random outcome: a diligent hunter would sometimes come home with no game, and a successful hunter might kill more game or a larger animal than he and his immediate family could consume. In this situation, food sharing would have been efficient as a form of insurance (Rubin 1982). Such sharing would be particularly useful if the marginal value to the recipient would be higher than the marginal cost to the donor (as, e.g., if the recipient would starve without the contribution and the donor would be in no present danger of starvation). The prisoner's dilemma occurs because sharing in such circumstances is efficient for both parties, but the party who first receives the transfer has an incentive to shirk and not reciprocate when he is the lucky hunter. Reciprocal altruism can make this policy privately as well as socially desirable, but in either case it would be an efficient form of altruism. Moreover, some of the rules that the group might efficiently enforce would be rules mandating sharing and reciprocation. We would expect humans to agree to share food, and food sharing in some circumstances is universal among humans (Brown 1991).

But it is important to note that such food sharing would be limited and would occur only in well-defined circumstances. Sharing creates risks of shirking of two sorts. One is not sharing when one is obliged to (the EEA equivalent of tax evasion). The other is shirking in food seeking or production in order to free ride on those who do produce and share (the EEA equivalent of welfare fraud). Kelly (1995, chap. 5) discusses these issues and shows that sharing, shirking, and detection of shirking all exist among hunter-gatherers. For example, hunters will sometimes consume game at the hunting camp to avoid sharing. Then food sharing would not be efficient if the recipient was lacking in food because of shirking. In the EEA, societies were relatively small, and monitoring to reduce shirking would often have been possible.

In addition to food sharing, assistance to the ill, hurt, or injured who would be expected to recover would have been an efficient form of altruism. This is because someone hurt or injured could, upon recovery, again become a useful member of the band. Injury was probably much more common in the EEA than today, because ways of making a living were more dangerous than occupations today; injury rates have been falling steadily in historic times and are lower now than even in our own recent past. Methods of production in the EEA, such as big game hunting with spears or hand axes, would have been particularly likely to cause injury, as would the individualistic modes of raiding and combat.

Even today many view medical care as a necessity and are opposed to treating medical care as simply another expenditure item (Daniels, Light, and Caplan 1996.) Such a taste for treating illness and medicine as being different could have evolved in the EEA. Moreover, the nature of care in the EEA was probably such that shirking in the form of consumption of additional care would not have been an alternative, since treatment possibilities would have been severely limited. This may be why people today think in terms of the amount of medical care needed without realizing that under today's conditions, the amount of care consumed can be increased if costs are reduced. This error has had important repercussions. For example, Medicare in the United States costs much more than the original proponents thought it would cost. This is because the founders of the system underestimated the extent to which consumption could increase in response to the reduced prices associated with the system (what economists call elasticity). Other proposals for universal health care must deal with this possibility.

None of these forms of efficient altruism would have supported providing food or other resources to a nonproductive individual who was not expected to become productive or to a shirker who was expected to continue to shirk. That is, efficient altruism might have supported transfers to a temporarily unlucky individual or to an individual who was temporarily incapacitated through illness or injury. But it would not have supported transfers to someone who would be expected to remain unproductive. Kin selection mechanisms might have induced family members to provide some support to elderly parents. However, even this support seems more common if the elderly parents have some estate they may leave to the children (Rubin, Kau, and Meeker 1979; Kronk 1999).

Continual transfers to nonproductive individuals would have been a drain on a band and would have reduced the ability to compete with other bands in the EEA. Providing transfers to shirkers will increase the number of noncooperators in the band; it will not control noncooperation. Of course, if the altruism associated with group selection would not have supported some policy, no other variant or form of altruism would have supported this behavior either since group selection is the most powerful mechanism known for generating altruism. Thus, we may have tastes for efficient altruism but not for indiscriminate altruism.

In chapter 1, I indicated that humans play different strategies in part because of frequency dependent selection. The same tendency applies to altruistic behavior—some people are more altruistic than others. We observe this in experimental situations, in which some individuals are more likely to cooperate than are others. This difference can explain laws mandating altruism. Altruists and recipients of altruism can form a coalition and pass laws mandating transfers from all, so nonaltruists are forced to contribute as well. However, if policies move toward inefficient altruism, then the altruistic members of the coalition may change their votes and pressures might limit altruism to only efficient forms.

On the other hand, shirking would also sometimes have been possible in the EEA, as it is today. Someone would have had an incentive to claim that he hunted unsuccessfully when in reality he had taken a nap just outside the living area. Thus, there would have been evolutionary pressure to share food with unsuccessful hunters and also to monitor for shirking. Some of the reasons given in the Inglehart, Basañex, and Moreno (1998) survey for doing voluntary work are of this type, including "moral obligation," "compassion for those in need" and "identify with people suffering" (tables V55–V65).

This argument has implications for the political analysis of transfers. Cosmides and Tooby (1992) discuss exactly this issue in the context of modern government provision of charity. They point out that advocates of charity or of using the political system for redistribution argue that the poor are poor because of random reasons, thus trying to enlist the mental module for sharing. (Advocates of less redistribution argue that the poor are lazy or are otherwise attempting to shirk, trying to engage the behavior associated with punishing of shirking.) In establishing other transfer programs, similar arguments are generally made. Rent seekers (those who use the political process to obtain wealth for themselves) usually do not argue that they deserve transfers simply because they have the political power to get them. Rather, they generally try to invoke evolved tastes for transfers to poorer group members based on altruism or fairness. (As discussed in chapter 7, they may well believe their own arguments, a form of self-deception.) Those seeking tariffs have an additional advantage: they can also appeal to group loyalty or patriotism with respect to other residents of the country, thus invoking the tribal or ethnic identity feelings that were important in the EEA. This may explain why international trade is often discussed in popular and political discourse using military analogies. (Of course, economists have known for two hundred years that such analogies are misguided.) Those opposed to immigration can appeal to the same forces.

Wax (2000) applies a similar model to the modern welfare system. She indicates that individuals support welfare and other transfers, but only to the deserving poor—those who are temporarily disadvantaged and are willing to try to earn money to leave the system. She also indicates that recent (1996) reforms in the welfare system are consistent with evolved preferences because these reforms emphasize work on the part of recipients. As a society, we are now spending more on assistance for the working poor than we have saved in welfare payments (Besherov and Germanis 2000), but this spending has met with less disapproval. This may be because the recipients are no longer viewed as shirkers. This may be an example of altruists changing their political behavior to favor only efficient altruism, as discussed previously. Recipients of transfers, or course, would favor both efficient and inefficient altruism.

The problem with charity in today's world is monitoring recipients. With small groups, it would have been quite straightforward to monitor recipients of transfers to be sure they were not shirking. In today's world, such monitoring is much more difficult because of the huge scale of society and the corresponding anonymity (Rubin 1982). Once transfers to the poor become available, many individuals attempt to obtain these transfers. One way to limit such shirking is to make transfers depend on characteristics of recipients. For example, to be eligible for Aid to Families with Dependent

Children, there must be children and, sometimes, no man in the house. However, individuals can be induced to conform to obtain the benefits. This is an example of rent seeking, and a classic form of exactly the sort of free riding that makes altruism difficult. Indeed, we may view all rent seeking as a form of free riding (Zywicki 2000.) Then expansion of government, which creates opportunities for rent seeking, itself makes free riding more likely and creates incentives for inefficient altruism.

This view of the nature of charity leads to some testable implications: First, individuals should be more willing to contribute to charity when the value to the recipients is greater than the cost to donors. For example, payoffs might be based on various thresholds with donors losing relatively little by falling below some high threshold and recipients gaining more by crossing a lower threshold. Second, individuals should also be more willing to contribute to charity when the lower income of the recipient is due to chance rather than to perceived shirking. This should be testable in terms of the story told regarding any income differences between the putative donor and the putative recipient in experiments measuring preferences for charity or redistribution. Third, there is a distinction between preferences for reducing the wealth of the most successful and preferences for benefiting the poor. (In chapter 4, I discuss the role of envy, and show that envy is aimed at reducing the wealth of the richest, rather than increasing the wealth of the poorest.) While some policies achieve both aims, some do not. It should be possible to distinguish between tastes for redistribution aimed at benefiting the poor and tastes aimed at reducing the wealth of the successful; both types of tastes are expected to exist but should be separate.

Evidence indicates that in hunter-gatherer societies, a so-called egalitarian ethic was present(Knauft 1991; Boehm 1993). This finding might seem to imply that we have inherited from our hunter-gatherer ancestors a desire for income transfers and equality of outcome. But careful reading of Boehm indicates that what was actually involved was a desire not to be dominated; this is consistent with Knauft's analysis, although Knauft is not as direct as Boehm. That is, individuals who expected to be the lower ranked members of any hierarchy banded together to prevent potential higher ranked members from actually forming a hierarchy. Evidence does not support the claim that anyone wanted to provide food or other goods to lower income individuals, nor does evidence show that potentially dominant members wanted to eliminate hierarchies. Rather, individuals wanted to avoid being dominated. Humans still have such preferences. For example, Buchanan and Tullock (1965) provide a theory of constitutions based on this desire. (I discuss this issue carefully in chapter 5.) But this is not the usual meaning of egalitarian in today's world.

Cheating: Examples

Humans are very smart and creative. We can figure out novel and productive ways of cooperating, such as forming large firms and states. And we can figure out novel ways of free riding, or cheating on the social contract. Some cheating occurs in small, face-to-face interactions, which is the sort of cheating typically studied in experimental situations. It is the sort that governs the prisoner's dilemma as the story is actually told. If we accept the arguments of Smith (1998) and Richerson and Boyd (1999), that there are two separate mechanisms for cooperation, then there are also two separate forms of cheating. So far, most analysis has focused on cheating in small, group face-to-face interactions.

Cheating also occurs in large impersonal settings and takes the form of appropriating the surplus created by large-scale cooperative activities. For one example, Betzig's (1986) discussion of despotism and the number of wives despots appropriated in several historic societies is a story of social cheating. More recently, economists and political scientists have begun examining rent seeking. Interest groups use rent seeking to derive special interest benefits from the government(Tullock 1967; Tollision 1982). Standard examples include subsidies or tax benefits to interest groups and laws preventing competition, such as occupational licenses or tariffs. Government sometimes facilitates collusion or cartelization of some market, as when it favors unions or agricultural crop-marketing associations. When such interest groups seek benefits, they typically use a public interest justification. That is, they do not simply assert that the have sufficient power to get benefits; rather they allege that the benefits will also benefit the public. As mentioned earlier, the theory of self-deception indicates that they may well believe these assertions themselves.

Some people also seek attempt to cheat without benefit of government assistance; examples include monopolization, discussed in chapter 4, and cartelization. A cartel exists when firms get together and collude to act as a monopoly. Interestingly, the possibility of free riding or cheating on the collusive agreement limits the ability of firms to collude. That is, if firms agree to charge high prices, then any one firm can make even more profits by charging a price slightly below the collusive price. In other words, the incentive for individual firms to cheat on the agreement that is aimed at social cheating limits the ability of firms to collude and jointly cheat on the social contract. The possibility of entry into the market is another limitation. If firms do successfully collude, others can enter the market and undercut their prices. For these reasons—internal cheating and entry—firms that want to engage in rent seeking find it useful to enlist the government in their efforts. The government has the power to prevent firms from cheating and also to prevent entry. The rent seeking insight is that government itself can be a powerful cheater in cooperative social games.

Bentham Yes, Marx and Rawls No

Three standard philosophical systems that discuss the optimal moral stance for society are utilitarianism (Bentham, 1781/1988), the Rawlsian system (Rawls 1971), and Marxism (Marx, 1888) or, more generally, socialism. Utilitarianism is roughly consistent with the form of altruism that would occur under group selection; the Rawls difference principle is not, nor is a communist or socialist system. My analysis here is uninformed by extended academic discussions of moral theory, which Posner (1999) indicates are not very fruitful anyway. E. O. Wilson (1998) argues that moral philosophy is not generally based on empirical biological knowledge. Alexander (1987, 145–165) discusses the relationship between moral philosophy and biology as viewed by moral philosophers. He concludes that in general moral philosophers have not paid adequate attention to biology and have not taken account of biological knowledge. Earlier, Westermarck (1932) had reached a similar conclusion regarding an earlier generation of moral philosophers. Singer (2000) is a work of moral philosophy based explicitly on Darwinian thinking. I refer to this work from time to time throughout the book. Singer does agree with my conclusions about Marxism (that it is inconsistent with evolved human preferences). He seems more disposed to Rawls, however, than I believe is consistent with these same preferences.

Bentham

Something approximating utilitarianism might well be consistent with efficient altruism. Utilitarianism argues for maximizing some function of individual utilities. Utility functions are related to (if not the same as) fitness: we get pleasure from those things that led to increased reproductive success in the EEA and pain from those things that hindered our ancestors' reproductive success (Johnson 1999). Thus, maximizing individual utilities would be the equivalent of maximizing group fitness in the EEA and would have been exactly the result of efficient altruism in a group selection process. This may be why utilitarianism has been a long lived and successful moral theory—it is a theory consistent with our evolved moral preferences.

Economists use an essentially utilitarian theory in welfare economics. But professional economists believe deeply in the impossibility of interpersonal comparisons of utility. That is, they are unwilling to compare utilities across persons. Modern economics, for example, is unable to say that transferring $100.00 from Bill Gates to a homeless person would increase total utility, because it is impossible to say that the increase in utility to the homeless person would be greater than the reduction in Mr. Gates' utility. The perspective here will at least partially solve the problem of interpersonal comparison. That is, if we take an evolutionary perspective, we can in some circumstances make interpersonal comparisons of utility. In particular, in those circumstances discussed previously, where the potential recipient of altruism can have his life saved by a relatively small transfer, then humans are often willing to make the transfer and are at least implicitly making an interpersonal comparison. In his Nobel acceptance speech and in other works, Amartya Sen (1999) argued for the importance of minimal nutritional and medical expenditures—a position consistent with the arguments made here, but derived from a totally different (and nonevolutionary) perspective. For example, Sen points out that no independent democracy has ever suffered a famine because the electorate will not allow it to occur.

Posner (1999) indicates that most academic criticisms of utilitarianism proceed by showing that the logical implications of utilitarianism are undesirable if the argument is carried to its logical extremes. But the argument discussed here is that utilitarianism is essentially the result of fitness maximizing preferences. In this reading, any implications of utilitarianism that conflict with fitness maximization for the relevant decision-making group are illegitimate extensions of the theory and should be ignored. This would apply, for example, to the sadist who gains great utility from torture, a standard counterexample for those opposed to utilitarianism. But such preferences would never have been fitness maximizing. Utilitarianism would perhaps be modified to maximization of utility in terms consistent with fitness maximization in the EEA. Of course, we now extend our altruistic preferences beyond the level of the group, so that some fitness maximizing behaviors—for example, massacre of males of neighboring tribes—would no longer be considered consistent with utilitarianism. J. Q. Wilson (1993) and Binmore (1998) among others discuss tendencies to expand the set of humans included within the group. Indeed, Westermarck (1932) makes a similar point. As we will see later, this is a useful but limited perspective for reducing intergroup conflict.

Rawls

Rawls argues that humans behind a veil of ignorance would choose an income distribution system based on the difference principal. Therefore, any inequality in income distribution can be justified only if it benefits the worst off individuals in society. By arguing that this is the rule individuals would choose behind the veil, Rawls is arguing that people are exceedingly risk averse. A huge amount has been written about Rawls's analysis, and I do not provide a thorough discussion. But note that this principle is not consistent with efficient altruism as defined earlier. A group adopting this principle would not have increased its average fitness.

Consider: Group A does not adopt Rawls's principle. All active adult males have a nutritional level of 2,500 calories per day except for one individual who obtains 2,000 and is somewhat malnourished (but not in danger of immediate death). Group B begins with the same distribution but adopts the difference principle and redistributes accordingly. There are two efficiency costs of redistribution. First is the deadweight loss due to reduction in effort as a result of taxation in the form of forced transfers. High-income or highly productive individuals will reduce their effort as they are forced to give part of their output to support others. Second, recipients will also increase their shirking as they receive transfers anyway. That is, low-income individuals will shirk in the expectation of receiving the benefits of redistribution. In other words, the difference principle will itself encourage free riding in any society that adopts it. These costs are so high that once the redistributive process has run its course, total income falls. As a result, at equilibrium, everyone in the group has a nutritional level of 2,100 calories per day. According to Rawls, Group B is morally superior to Group A, because the poorest person in B has 2,100 calories and the poorest in A has only 2,000. Rawls's argument is that a randomly chosen person who does not know his position in society would choose the distribution in Group B rather than that in Group A.

But though Rawls and his disciples may prefer this outcome, it is not the outcome that would be selected in the EEA. Rather, members of Group A would translate their extra nutrition into increased strength or more hand axes and likely massacre the males (at least) in Group B. Or, in a more benign scenario, many productive members of Group B would migrate to Group A. In other words, the cost of Rawls's policy is so large that it would not meet the Sober and Wilson criterion for being the outcome of a group selection process, and so it is unlikely that humans have been selected to desire Rawls' preferred outcome. That is, members of Group B would not be our ancestors, and their set of tastes would not have survived.

We are descended from ancestors with tastes more like those of Group A. This is particularly relevant for studying Rawls, since his argument is that people would choose the difference principle behind the veil of ignorance, essentially a statement about utility functions. But the argument here is that the utility functions that Rawls postulated could not have evolved, and so we cannot have them.[2] Alexander (1987) has a discussion of Rawls' wherein he confines his analysis to a discussion of the veil of ignorance, arguing that this may be an appropriate construction in a society in which individuals are less able to foresee the positions of their descendants. He does not discuss in detail the hypothesized outcome of the process. Wilson (1998, 249) does indicate that Rawls "offered no evidence that justice-as-fairness is consistent with human nature." Nonetheless, the construct of the veil of ignorance is a useful one, as long as the analyst using this tool correctly chooses preferences.

Marx

A socialist system divorces returns from inputs. ("From each according to his abilities, to each according to his needs.") It also eliminates the link between ownership and return, thus leading to inefficient resource use. As we have seen in recent years, such systems are inefficient and lead to reductions in output. Again, they encourage, rather than contain, free riding. As such, in a highly competitive environment such as the EEA, they could not survive. Indeed, extreme socialist regimes apparently cannot survive even in our relatively more relaxed environment. Moreover, tastes for socialism would not have evolved, because those with such tastes would not have done well. Bailey (1992) has shown that efficient definitions of property rights are virtually universal in hunter-gatherers. Brown (1991) indicates that private property and inheritance are universal among humans. Boehm (1997b, S117) argues that socialism is a form of altruism but one that is "so very difficult to execute." The difficulty is associated with its inefficiency and more generally with evolved preferences that are inconsistent with this social arrangement. (I discuss Marxism more fully in chapter 4.)

Fairness

Humans have a preference for fairness. Trivers (1971) indicates that we feel moralistic aggression if we believe that we have been treated unfairly; this is part of the mechanism for policing reciprocal altruism. But the concept of fairness is not well defined.

Experimental evidence indicates that people are concerned about fairness (see, e.g., Fehr and Schmidt 2002.) Much of this evidence comes from an experiment called the ultimatum game. In this game, a certain sum of money (say, $10) is to be divided between two individuals, the Proposer and the Responder. The Proposer makes an offer—for example, $6.00 for the Proposer, $4.00 for the Responder. If the Responder accepts, the money is divided as proposed. If the Responder does not accept, neither person gets anything. If the Responder is a rational, selfish person (say, a neoclassical economist), he should accept any offer of more than zero. If the Proposer knows this and is also a rational, self-interested person, he should make an offer like $9.90 for the Proposer, $.10 for the Responder.[3] In fact, however, most Proposers offer more than this—usually about 20 percent of the value ($2.00 in this example) and if less is offered, the Responder often (40–60 percent of the time) refuses, so both get nothing.

The simplest explanation is that Responders reject offers they perceive as being unfair and Proposers, anticipating this, do not make offers that they expect will be turned down. These results are quite robust and persist through many experimental treatments. They do vary across societies, however, with the amount offered by the Proposer being positively related to the use of the market and also to the payoff to cooperation in the society (Henrich et al. 2001). Because of experimental results such as these and results dealing with public goods, discussed later, many economists (e.g., Gintis 2000a, 2000b) have postulated that much human social behavior can be explained in terms of strong reciprocity, which is defined as willingness to cooperate with others and to punish noncooperators.

Lakoff (1996, 60–61) has an interesting categorization of various views of fairness. He suggests at least 10 ways in which the term can be used:

1. Equality of distribution (one child, one cookie)
2. Equality of opportunity (one person, one raffle ticket)
3. Procedural distribution (playing by the rules determines what you get)
4. Rights-based fairness (you get what you have a right to)
5. Need-based fairness (the more you need, the more you have a right to)
6. Scalar distribution (the more you work, the more you get)
7. Contractual distribution (you get what you agree to)
8. Equal distribution of responsibility (we share the burden equally)
9. Scalar distribution of responsibilities (the greater your abilities, the greater your responsibilities)
10. Equal distribution of power (one person, one vote)

Lakoff does not furnish any basis for this list of meanings or intuitions. (Carroll [1999] has recently provided a general critique of Lakoff's work indicating that the work is not grounded in evolutionary theory and would be improved if it were.) But it is easy to see that Lakoff's categories are consistent with the evolutionary explanation for altruism proposed here. That is, under proper circumstances many of Lakoff's categories could be associated with a form of efficient altruism. (Of course, under different circumstances, many of these categories could be mutually inconsistent, which is why economists generally avoid the use of the concept of fairness.) Moreover, the actions associated with many of these concepts are universal among humans (Brown 1991). I briefly consider each.

EQUALITY OF DISTRIBUTION

This is consistent with food sharing, which, as discussed previously, is sometimes an efficient form of altruism. Some food sharing, although not equality of distribution, is universal among humans (Brown 1991). Full equality, however, is not be an efficient form of altruism, because it leads to excessive shirking.

EQUALITY OF OPPORTUNITY

This would probably have been the norm in the EEA with respect to wealth. It would also have served to reduce the power of dominant individuals by reducing their ability to pass on their success to their children. Among mobile hunter-gatherers and their predecessors, hereditary wealth would have been minimal (since the mobile nature of the society meant that wealth in general was minimal), so the tendency would have been toward de facto equality of opportunity. Nonetheless, evidence indicates that children of dominant individuals do well in human and even chimpanzee societies. Today the term is used loosely: no one (in America at least) advocates true equality of opportunity, since this means removing all children from their parents at birth and raising them communally. Otherwise, differential parental abilities in child raising and different levels of parental wealth will effect children. This is consistent with patterns in the EEA. Additionally, children can of course inherit wealth in our society.

PROCEDURAL DISTRIBUTION

At least to some extent, playing by the rules is important. Brown (1991) indicates that rules are universal among humans and that sanctions for violations are also universal. If the rules are themselves aimed at forcing cooperation in prisoner's dilemmas and at enforcing reciprocity as Sober and Wilson indicate, then tying income distribution to obeying rules is efficient.

RIGHTS-BASED FAIRNESS

If rights are defined efficiently, this form of fairness is also efficient. Private property rights are of course an efficient way to define rights (e.g., Pipes 1999). Brown (1991) does indicate that ownership of property and inheritance of property are universal among humans. Bailey (1992) shows that among hunter-gatherers, property rights are generally defined efficiently. To the extent that rights-based fairness is a form of property rights based on fairness, we can understand this notion and associated preferences. This form of fairness is also consistent with the bourgeois strategy in the hawk-dove game.

NEED-BASED FAIRNESS

This is consistent with an insurance model of sharing. Those who have a shortfall in a period receive transfers. This keeps them alive as potential fighters when needed (a group selection argument). This form of behavior may also be part of a process of reciprocal altruism. Of course, for this form of altruism to be efficient, some monitoring to reduce shirking is needed. Moreover, need would be carefully circumscribed, perhaps including only being at risk of or close to immediate death.

SCALAR DISTRIBUTION

This is an efficient principle of distribution because it provides incentives for work. In most hunter-gatherer societies, the hunter who kills the game has rights to distribute it, subject to various social norms. If this were the pattern in the EEA (and chimpanzees also behave in this way, suggesting that it is an evolutionary old behavior, predating the hominid line), it might be an evolved principle of distribution and a form of efficient altruism.

CONTRACTUAL DISTRIBUTION

This is equivalent to reciprocal altruism, an agreement to return a gift, and is an efficient method for solving intertemporal prisoner's dilemmas. A norm of reciprocity is universal among humans (Brown 1991). Of course, today we have third party (court) enforcement of contracts, so possibilities for reciprocity are greatly expanded.

EQUAL DISTRIBUTION OF RESPONSIBILITY

This principle serves to constrain dominant individuals in society and to create more equality. It is consistent with Boehm's arguments about the reverse dominance hierarchy.

SCALAR DISTRIBUTION OF RESPONSIBILITIES

This seems to be an important principle in hunter-gatherers. Better, more skilled hunters do catch more game even though they are obligated to share much of their catch with others (Kelly 1995). Therefore, the behavior is itself somewhat of a puzzle. Kelly provides some hypotheses to explain the behavior, but ultimately the reason for this seems to be an enigma. Such behavior is an efficient form of altruism and could be consistent with a group selection model. Of course, today, in a market economy, rewards are relatively closely related to effort.

EQUAL DISTRIBUTION OF POWER

As Boehm (1993) and Knauft (1991) have argued, in small hunter-gatherer bands, which were probably the most important human and hominid population structure in the EEA and for most of the existence of humans and our immediate ancestors, there is something like a reverse dominance hierarchy. This is interpreted to mean that no individual is dominant and decisions are jointly made. Brown (1991, 138) indicates that no humans have perfect democracy and none have complete autocracy, so "they always have a de facto oligarchy." This principle would be associated with reducing the power of dominant individuals and with Boehm's reverse dominance hierarchy. While many human societies have been centrally run for the benefit of dominant individuals (Betzig 1986), relatively democratic or egalitarian societies are preferred, when possible. (This issue is discussed more fully in chapter 5.)

Thus, we see that at least five of Lakoff's principles are associated with efficiency: procedural distribution, rights-based fairness, scalar distribution, contractual distribution, and scalar distribution of responsibilities. Two principles are associated with insurance: equality of distribution and need-based fairness. The other three principles, equality of opportunity, equal distribution of responsibility, and equal distribution of power, reduce the power of dominants. Our intuitions about fairness seem to be consistent with evolution of tastes in the directions discussed in this book.

This list and the discussion may seem trivial. But this is because the notions are so intuitive that it is not easy to see how alternatives could exist. Consider some potential notions of fairness or morality that no one would endorse:

> Take from the poor and give to the rich. Cheaters who cheat fellow group members deserve as much as they can get away with. Promising to return a favor and then reneging is moral. Compensation should be inversely related to effort. Good hunters should stay at home and poor hunters should have the responsibility for bringing in game. Each person should do that at which he is worst (has a comparative disadvantage). The least competent person, perhaps a child, should make decisions for the group. Arbitrary dictatorship is a moral form of government. No good deed should go unpunished.4

Obviously, these are silly principles and clearly not intuitively fair in any sense. But that is the point: they do not seem fair or desirable, and they would have been associated with group and individual death in the EEA. If any of our predecessors had beliefs such as these, they did not survive to be our ancestors. We have those intuitive notions of fairness that led to survival, and we still view these behaviors as fair.

Rules Governing within Group Behavior: The Origins of Law

If groups can behave efficiently, they can prosper relative to competing groups. Various types of rules favor efficiency within groups. These can be enforced through any of the mechanisms discussed earlier (use of the bourgeois strategy, asymmetric interactions, first- and second-degree punishment). These rules are easy to understand and not surprising. They are essentially the basic legal rules defining property rights, enforcing contract, and outlawing certain forms of violence—the core of the common law. They are integral to any society, and some variant of these rules is universal among humans. Essentially, these rules are institutionalized methods to contain free riding.

Property

Consider property. All humans define themselves in terms of a territory and have concepts of individual and group property (Brown 1991). Property rights are efficient and lead to increased wealth, however wealth may be defined for a society. Hunter-gatherers have defined efficient property rights (Bailey 1992). For example, while territory is shared for hunting, agricultural land is privately owned and crops belong to whoever planted or tended them. Property rights can change seasonally, depending on whether land is being used for crops or for hunting. Property is sometimes considered a result of the bourgeois strategy in the hawk-dove game as defined by Maynard Smith (1982). This is the strategy, "If you are first to get to a resource, play hawk; if not, play dove." That is, whoever gets to a resource first owns it. This strategy dominates either a pure hawk strategy or a pure dove strategy. While chimpanzees do not have much to own, they do seem to have implicit property rights: those who capture game (usually monkeys) have at least some rights in the prey (de Waal 1996).

Property rights are often considered similar to territoriality in birds and animals. In many species, individuals defend territories, and others honor these territories. This may be the result of all parties playing the bourgeois strategy. Territoriality has also been observed in chimpanzees, but the borders of territories are often contested. As the range of a band shrinks, animals become physically lighter in weight and fertility declines (Pusey 2001).

Thus, territory seems to play an important economic role in the lives of chimpanzees.

Contract

Humans also enforce contracts. They recognize and enforce promises and engage in trade and exchange (Brown 1991). Trivers (1971) rediscovered contract and called it reciprocal altruism. He argued that humans feel injured and engage in moralistic aggression if promises are broken. McGuire (1992) discusses the evolved mechanisms for generating moralistic aggression. Charlesworth (1992) shows that similar mechanisms operate in children. De Waal (1996) discusses chimpanzees. He shows that after a hunt, the hunter who ultimately has possession of the game rewards participants according to their level participation. Since the hunt is over, this is a sort of contractual compensation for completed actions. Apparently baboons do not engage is such sharing and, consequently, sometimes miss opportunities for acquiring meat.

While the amount of exchange is limited in primitive societies, as discussed in chapter 1, some exchange does occur. An important form of exchange—and one that does require some governance—is intertemporal exchange. If I have a lot of meat today and you have none, I may give you some in the expectation that you will reciprocate when the situation is reversed. But because the exchange is not simultaneous, some enforcement method is required or else free riding is possible. Even today, in societies with little rule enforcement, reciprocation is important, as is trade with ethnic relatives (Landa 1981; Rubin 1994). Smith (1998) discusses this issue and suggests that we are selected to engage in impersonal exchange through markets and also personalized exchange with known individuals as a form of reciprocity. This notion of a two-stage process is similar to that which Richerson and Boyd (1999) proposed. In this sense, the issue of the primacy of political or economic institutions is not a real issue; humans as humans have always had both sets of institutions.

Crime

Something like criminal law is found among all humans, again aimed at limiting noncooperation. All humans limit within-group violence and punish violations (Brown 1991). All societies forbid many types of rape (Edgerton 1992). In primitive societies, these constraints are weaker than in advanced societies; death rates by homicide are much higher among hunter-gatherers than even in the most dangerous contemporary western societies. Such punishment is never fully effective; violence and crime are universal. Daly and Wilson (1988) have shown a biological basis for violence. (This does not mean that there is an innate drive for violence. It means that violence is a biologically useful strategy in some circumstances, and understanding these circumstances helps understand patterns of violence.) Human societies try to reduce the level of such violence. Some societies are better at reducing violence than others, but all try and none perfectly succeed. Indeed, de Waal shows that this mediating behavior also exists in chimpanzees and is not perfect there either. Keeley (1996) points out that our modern society is better at limiting violence than are more primitive societies.

Rules

These rules of property, contract, and crime govern behavior within the group to some extent. It is possible to find precursors to property rights and to enforcement of quasi-contractual promises among chimpanzees; de Waal (1996), Ridley (1996), Grady and McGuire (1997, 1999), and McGinnis (1997), among recent authors, discuss such rules and rule-like behavior. Cosmides and Tooby (1992) present evidence that humans are better at making analytical judgments when the issue is the honoring of a commitment. This indicates that cheating on agreements has long been a feature of human behavior, and so we have evolved sophisticated methods of detecting such cheating. The prisoner's dilemma with respect to honoring of contracts and of respecting property rights is clear. Numerous institutions exist even in primitive societies for solving or mediating these dilemmas (e.g., Hirshleifer 1977, 1980; Posner 1980; Epstein 1980, 1989, 1990; Bailey 1992). All humans have sanctions for violations of rules (Brown 1991). Ridley (1996) provides a nice discussion of such mechanisms and relates the economic property rights literature to the biological literature.

These rules are not created, even at the most primitive level. That is, humans came into being with some rules governing property, contract (honoring commitments), and in-group violence already in place. As societies and transactions became more complex, these rules multiplied and became more complex, and enforcement mechanisms also evolved—into our present hyper-elaborate legal systems. But neither proto-humans nor humans lived in a time without rules or in a time when they were forced to write such rules de novo. In this view, there is no sharp difference between social norms and law; rather, all rules begin as norms of some sort and as complexity grows, some norms become enforced as laws. Moreover, the scope of rules consistent with evolved human preferences and abilities is not limitless, and it is possible to write a logical set of rules that in fact are internally consistent but inconsistent with human nature and unworkable. An example is Russia and Eastern Europe under communism, where great misery resulted from the assumption that human nature is easily malleable (Pipes 1990).

Stories and myths about the origin of laws seem common (the Code of Hammurabi, Moses on the mountain) because humans seem unable to grasp the notion of a fully evolved system with no planned elements. (Indeed, it was not until 1776 that the notion of an unplanned but functioning economy was understood, and even now, many do not comprehend it.) This is consistent with Boyer's (2001) argument that humans experience social life but do not understand it very well. Myths about the origin of law may lead to increased obedience. The common law of England, so beloved of economists (including myself; see Rubin 1977), is merely one stage in this evolutionary process. The common law or any other existing body of law, if it could be traced back sufficiently far, would turn out to be directly descended from the evolved predilections for rule-based behavior. Contemporary Russia is experiencing difficulty in creating a body of law because of the discontinuity in legal change caused by communism and communist law. It appears to be very difficult to create a functioning body of law de novo, with no evolutionary or historical component. Poland, Hungary, and the Czech Republic have been more successful, because they have been able to adapt previous bodies of law, continuing an evolutionary process (Rubin 1994).

One important issue in law and economics and elsewhere is the origin of property rights. For example, in discussions of wealth maximization, Posner (1979) and his critics (e.g., Keenan 1981) address the origin of property rights. Posner advocates using wealth maximization as a normative principle. However, wealth maximization is only defined if there is a set of prices and prices imply some distribution of wealth in society. Thus, the criticism is that Posner's suggestion is circular—wealth maximization is only defined with respect to a distribution of wealth, but the goal of wealth maximization is to determine how to allocate wealth (broadly defined.) The error on both sides in this line of analysis seems to be in the assumption that any distribution of wealth is arbitrary and a product of some decision or allocation process. But if wealth distributions are basically given for evolved or historic reasons, they are not arbitrary and it is possible to start from the existing distribution. Of course, the existing distribution is not immutable, nor is it in any useful sense morally correct, but on the other hand it is not arbitrary, and there is no reason why it is improper to start with this distribution in performing the wealth maximization analysis.

Public Goods

A major puzzle facing economists is the extent to which humans voluntarily contribute to public goods. The standard model of utility maximizing relatively selfish individuals (which is the same as the individual selection model discussed previously and of the cheater strategy in the prisoner's dilemma) that economists use cannot explain observed behavior with respect to public goods. Of course, most public goods are financed through taxation, which is not voluntary. But both experimental and real world evidence of voluntary contributions is inconsistent with conventional economic theory. Ledyard (1995) and Fehr and Schmidt (2002) summarize the experimental evidence. Hoffman, McCabe, and Smith (1998) also discuss this evidence, from a strictly evolutionary viewpoint. The evidence consists of various sophisticated replications of versions of the prisoner's dilemma or related games, where playing the cooperate strategy is viewed as contributing to a public good. In addition to a willingness of many to cooperate, it is also found that, if the experimental treatment allows, some persons are willing to give up resources in order to punish those who do not cooperate.

In nonexperimental settings, we may identify two separate types of public goods. First, people contribute time and money to charity. Inglehart et al. (1998) indicate that in their sample, 60 percent of respondents belong to some voluntary organization and 50 percent perform unpaid voluntary work for some organization. Moreover, people vote to redistribute money to others. While the income distribution may be viewed as a public good (Hochman and Rodgers 1969), there is no explanation as to why some individuals would include the incomes of other, nonrelated individuals as elements in their utility functions. Donors to charity do not themselves benefit in terms of income from the provision of the public good. (I discussed motives for charitable contributions earlier.)

Second, people contribute toward some group benefit, as through voting, volunteering for military service during a war, or providing voluntary contributions for goods such as public television. In such situations, donors directly benefit because they are members of the group consuming the public good, but of course free rider problems still exist. In the Inglehart et al. (1998) sample, people belong to and contribute to both types of organizations. Biologists also distinguish between altruistic behavior that benefits the altruist directly and that which benefits other members but not the altruist (Bradley 1999).

In experimental situations, willingness to contribute is often tested in the context of the prisoner's dilemma, where both parties gain from cooperation. While the experimental results are somewhat confused, generally more is contributed than would be consistent with strong economic models based on self-interest. In the Inglehart et al. (1998) survey, reasons for voluntary work include "contribution to local community" and "for social or political change," (tables V55–V65) both of which seem to benefit the contributor as well as the recipients. Here the group nature of selection pressures is relevant. The biological theory predicts that individuals would contribute to public goods more if they were induced to behave as if they were members of particular groups.

Hoffman et al (1998), Ledyard (1995), and Thaler (1992) discuss some evidence consistent with this argument. In particular, some experiments have found that communication between subjects increases cooperation, even when the communication is not directly related to the experiment. This is only so if the public good will benefit those undertaking the discussions, however. Thus, the theory predicting that contributions will be increased if people feel that they are members of some group are borne out. However, further experimental examination of this hypothesis would be useful. A testable prediction from the evolutionary theory is that individuals should cooperate more fully in prisoner's dilemma and other public goods environments if they can be made to feel as if they are members of a group or team. For example, subjects might be told that they are competing with another group and the group with the highest payoff will win some additional prize.

This analysis may shed some light on the paradox of voting. Any one vote has a vanishingly small probability of affecting the final outcome of an election. Voting is costly. Therefore, theory would predict that (almost) no one would vote. And yet people do vote in significant numbers. Perhaps voters identify with parties and behave as if voting is a way of contributing to a public good for the group. If parties can induce voters to participate in order to benefit the party being viewed as a group with voters as members, then citizens may be more likely to vote. The public good toward which voters are contributing is not the improvement in public policy, but rather the strength and fortunes of the party viewed as a group of which the voter is a member. Similarly, if citizens belong to interest groups and the interest groups can induce them to view voting as a contribution to the group, this may also explain some voting. Nelson (1994) makes a similar point with respect to voting by members of ethnic groups. He argues that political participation establishes bonds with relatives and friends.

Summary

In a theory of group selection, those groups or bands of humans with more cooperative preferences would have thrived relative to less cooperative groups, and so we would be descended from cooperative individuals. The problem with this argument is that free riding would pay within groups, and so we should expect humans to be fully selfish. This can be modeled as a prisoner's dilemma game, where cooperation is jointly rational, but each individuals has an incentive to behave noncooperatively. The puzzle of altruism is that some animal species, particularly humans, display significant cooperation.

Several theories have been proposed to explain cooperation, or altruism. One is kin selection: it pays to be cooperative to relatives because we share genes with them, and a cooperative act can benefit copies of our genes in relatives. A second theory is reciprocal altruism: we provide benefits to acquaintances today in the expectation that they will return the favor in the future. Another theory is that we play the bourgeois strategy and whoever gets to a resource first has de facto ownership; this is viewed by some as the basis of property rights. There are also theories based on punishment for non-altruistic behavior. There are theories that individuals associate with others who are altruistic; in this theory, humans have incentives to appear altruistic even if they are not and also to detect such deception. One valuable tool in this arms race is self-deception; we are better at deceiving others if we first deceive ourselves, which may explain why interest groups that seek benefits from the government often believe that these benefits are truly in the public interest.

Recently, Sober and Wilson have proposed a variant of group selection as an explanation of human altruism. This theory is controversial, but if correct, it would provide a basis for some limited forms of altruism; that is, humans could not be more altruistic than would be consistent with this theory. This theory, which I call efficient altruism, would explain transfers to individuals who are unlucky or injured to keep them alive for the future. It would not explain tastes for transfers to shirkers or to those who are not expected to be able to contribute to the group in the future, and so would explain monitoring to be sure that disadvantaged individuals are not shirking. This theory is consistent with utilitarianism, but not with Marxism or with Rawls's difference principle. It also helps explain intuitive notions of fairness.

Groups of humans have rules governing within group behavior. These are essentially some variant of property rights, of contract or promise enforcement, and criminal law, aimed at reducing within-group violence. Humans seem to voluntarily contribute more to public goods than economic theory would indicate. These include charity and also contributions to what I call group goods such as voting.

In contemporary United States society, we find a well-developed welfare system. There was much discontent with the system; however, recent reforms have emphasized work requirements for welfare recipients. There seems to be less unhappiness with this system; this is consistent with evolved preferences for transfers under limited circumstances, when shirking can be controlled.

CHAPTER 4

Envy

In chapter 3, we saw that humans might have a taste or preference for redistribution toward the poor or ill and that such tastes could easily have evolved. This is one motive for income redistribution. Another motive is envy. Envy may have an evolutionary basis as well. In this chapter, I discuss the basis and effects of envy.

Attitudes about the Rich

One way to finance redistribution toward the poor is by the use of taxes, and particularly by progressive taxes—taxes that increase as a percentage of income as income increases. But there is another motive for progressive taxes and other policies that sometimes penalize the more successful members of society. Many people feel envy toward the relatively rich. In economic terms, for some people, the wealth of others enters negatively into utility functions. Thus, many persons seem to feel that the wealthy are morally unworthy in some sense. This feeling is linked to a belief that the only way to accumulate wealth is to take it from others, so the wealthy have acquired their riches at the expense of the less wealthy. Boyer (2001) points out than feelings of envy are often based on the possibility that one acquired wealth through social cheating. He indicates that such feelings are often associated with a fear of witchcraft or of curses such as the evil eye. In contrast, economists generally agree that, in most cases, in market economies such as those in the United States and Western Europe, the most efficient and most common way to accumulate wealth is to provide some productive service for others. Thus, in most cases intuitions about the rich are incorrect, in that the wealthy have not in general accumulated their riches through exploitation.

It is easy, however, to see how a basis for such incorrect attitudes has evolved. Consider first the accumulation of wealth in a primitive society. In such societies, income can fluctuate widely for essentially random reasons. (This is not to deny that some hunters or gatherers will be more skillful and harder working than others.) In such circumstances, individuals will be subject to significant variations in income for no reason. Because humans are in general risk averse (that is, they desire to avoid variations in incomes) for evolutionary reasons, as discussed later in this chapter, there would be a desire to avoid such fluctuations. Therefore, sharing food in such societies would be desirable. In fact, as discussed in chapter 3, many institutions of primitive society can be understood as methods of acquiring insurance (i.e., of sharing the risks of uncertain streams of income). If someone is wealthy in such circumstances, probably he is not sharing his income in the socially prescribed manner; that is, the way to accumulate wealth in primitive societies is to shirk one's duty to share and instead to hoard an unfair share of resources.

Risk

Risk aversion means avoiding a fair gamble. If I can bet $50 with a one-third chance of winning $150 and a two-thirds chance of losing the entire $50, this is a fair gamble, because the expected value of the gamble is exactly equal to the price, $50. A risk neutral person would be indifferent between keeping the $50 and taking the gamble. A risk averse person would prefer the $50 to the gamble and would not voluntarily participate, unless he could pay less than $50—say, $45—for a chance in the gamble. A risk seeker would pay even more than $50—say $60—for the gamble. Most of the time, most people act in a risk averse manner, as when we buy insurance, although in some circumstances—say, playing the lottery—people act as risk seekers.

The evolutionary explanation for risk aversion is straightforward: in a subsistence economy such as was probably the norm in the EEA, any significant reduction in income or wealth (food) leads to death. Moreover, in such a society, little if any storable wealth existed; if a hunter killed two deer today, this would probably not have increased his consumption, because the second would spoil before it could be eaten. Therefore, gains in wealth would be of relatively low value, while reductions in wealth would lead to starvation. Thus, humans and other animals are selected to avoid risk and attempt to garner enough resources to survive in the short run. Consequently, at least with respect to goods, risk aversion—the avoidance of even fair gambles—would have been selected for.

For example: One of our prehuman ancestors was in a fruit tree with enough food to survive for another day. There is a 50 percent or greater chance that another tree that is out of sight will have at least twice as much food. Thus, a decision to search for the other tree is at least a fair gamble and perhaps a better than fair gamble (if the chance is higher than 50 percent of finding the other tree or there is more than twice as much fruit in the other tree). Nonetheless, it would not have paid to have left the current tree and searched for the better tree. The gains in utility or fitness from the increased fruit would not outweigh the potential losses, since these include starvation. If our predecessor bet right and found much more food, the fitness gains would have been small. If he bet wrong and did not find more food, he might have starved and left no descendants. (He would not, in fact, have been our ancestor.) If he did not take the gamble (remained in the first tree), he would surely have survived. Therefore, we are descended from those individuals who avoided such gambles. This is the evolutionary basis for risk aversion and can explain why people today buy insurance and why the stock market devalues risky or volatile stocks. It cannot explain why people gamble and purchase lottery tickets. On the other hand, in some situations risk taking would pay for gains in status, as discussed later in this chapter.

Income in the EEA and Today

In addition, as discussed in chapter 1, the EEA may have approximated a zero-sum society. There were few possibilities of gains from trade or from devising efficient methods of production or exchange, or from productive investment. If so, we may have evolved attitudes consistent with such an environment. But in a zero-sum environment, the only way to become rich is to take from others. That is, if possibilities of increasing wealth by increasing production are not available, the only road to wealth is through predation or theft. This is no longer the case, but many of our attitudes seem consistent with this view of the world.

In modern societies, one effective way to become wealthy is to develop some new idea. In primitive or evolutionary societies, this method of enrichment is largely lacking. One reason is that such societies afford little privacy (perhaps a reaction to high information costs) and, therefore, little possibility of appropriating the income from new ideas (Posner 1980). If the value of an innovation cannot be appropriated, spending the resources needed to innovate is of little value. Thus, a person who was wealthy in a primitive society was unlikely to have achieved this wealth through innovation.

Indeed, the rate of technological innovation in primitive societies seems remarkably—indeed, almost inconceivably—slow by current standards. (This paragraph is based on Gowlett 1992.) For example, the Acheulean hand axe tradition lasted for more than one million years in Africa, Asia, and Europe. In the Upper Paleolithic, about forty thousand years ago, major technological change is defined as occurring when a change in stone techniques occurred over a few thousand years. Even more recently, the Gravettian tradition in Europe lasted from about twenty seven thousand to about twelve thousand years ago. This slow rate of technological innovation may have been due to relatively less intelligent prehuman ancestors in the early periods. In more recent periods, low levels of population and hence fewer individuals to create new technologies are the cause (J. Simon 1981/ 1996; Kremer 1993; Jones 2001). Edgerton (1992) discusses the slow rate of technological change in more contemporary societies. But whatever the source, the point is that wealth in the evolutionary environment would have not been the result of inventive activity.

If the situation in which the only way to accumulate wealth is to hoard or to predate against others was common to humans and to our ancestral species for long periods during evolutionary times, it is possible that the basis for feelings about the immorality of wealth actually became genetically established. That is, if the only way to become wealthy during most of human evolution was to shirk, selection pressure for disliking the wealthy would have occurred. This might be another implication of Trivers's concept of moralistic aggression—that is, it might have paid for our ancestors to feel moralistic aggression toward the wealthy, because the wealthy would have always been those who did not fulfill their reciprocally altruistic obligations. Of course, as with many other tastes described in this book, I do not claim that everyone will share such feelings—only that at least some humans will have these beliefs or at least be easily able to learn them.

Evidence indicates that some people do intuitively feel resentment toward the rich. Such attitudes are a common part of religion; consider the Christian statement that "It is easier for a camel to go through the eye of a needle, than for a rich man to enter the kingdom of God" (Matt. 19:24). For another example, consider how television depicts the wealthy. Stein (1979) demonstrates that a disproportionate number of criminals on television shows are businessmen and that businessmen are generally depicted as evil, immoral, and dishonest. Stein attributes this to the attitudes of those in charge of making television shows. One would expect, however, that people who made television shows that contained messages conflicting with the beliefs of the audience would not be successful. Moreover, the idea of businessman as crook seems to be an attitude older than modern television—for example, the evil businessmen out to steal land from poor ranchers and farmers seem to have been a staple of western movies for some time. Additionally, our tastes in stories likely evolved so that we may enjoy those stories which have plot elements that have had evolutionary usefulness, as discussed in chapter 7 (Scalise Sugiyama 2001). This implies that we might in some sense be selected to believe that the rich are evil, and that successful religions, television shows, and novels simply portray those plots which humans will find pleasant. Inglehart, Basañex, and Moreno (1998) report that 21 percent of their respondents believe that "people can only accumulate wealth at the expense of others" (table V256), an attitude consistent with a zero-sum world. Similarly, 26 percent of their sample believe that "incomes should be made more equal" (table V250).

Wealth in Modern Societies

The argument here is that the feelings we have about morality are sometimes counterproductive. To demonstrate that this is so, we must now consider the actual methods of accumulating wealth. Specifically, we must ask if the accumulation of wealth is actually in general socially counterproductive; that is, perhaps our feelings are correct and most of the wealthy have in some sense exploited their riches from others. To anticipate, the argument economists generally give is that most wealth accumulation in contemporary capitalist societies is the result of socially useful processes, and, therefore, attitudes leading to policies that penalize such behavior are in general counterproductive.

We may identify four ways in which wealth can be accumulated: theft, monopolization, productivity increases (including technological or managerial innovations), and simple saving and investment. The first is counterproductive and should be penalized or viewed with condemnation. Monopolization may sometimes be undesirable, but not always. The third and fourth methods are productive and should be applauded. Most wealthy persons have either accumulated their wealth by increased productivity or savings or inherited from those who have done so, and thus most wealthy persons have created substantial benefits for society or inherited from those who have created such benefits. To the extent that we assume that wealth has been accumulated by either of the first two methods, we are allowing our intuitions to rule incorrectly.

Crime

Consider first crime as a method of wealth accumulation. Outright theft does of course occur, as does fraud. Both of these methods of wealth accumulation are socially counterproductive. Resources are wasted in criminal behavior itself and in attempts to discourage the behavior, such as locks and police, and criminal behavior itself adds nothing to social wealth. It is unlikely, however, that theft or its variants in any sense accounts for any significant fraction of the wealth of the wealthy. Even in organized crime, theft is probably not a significant explanation for the accumulation of wealth. Rather, organized crime makes its income by providing goods and services that consumers want to buy but which are outlawed (Rubin 1973), such as drugs and gambling. (Whether the same results hold in contemporary Russia is a more difficult question, but I am dealing here with the United States and Western Europe.) Moreover, as mentioned above, methods of contract enforcement available to criminals cannot support very large organizations, which limits the ability of criminals to accumulate wealth.

Monopolization

Monopolization can secure above normal incomes. However, the extent to which this has been the major source of high incomes is probably greatly overestimated by noneconomists. One estimate is that if concentration in manufacturing (a measure of monopoly power) were reduced so that all industries were perfectly competitive, the income of the richest 20 percent of Americans would fall by 6 percent and the income of the lowest income class would increase by 3 percent (Powell 1987). Of course, much concentration is due to factors such as economies of scale and is unrelated to monopoly power, so this analysis greatly overstates the extent of harmful monopoly power in the United States. If we examine the twenty-five wealthiest families in the Forbes list for 1988, only six derived their wealth from industries that are arguably monopolistic, and these are mainly newspapers (often local monopolies) and broadcasting, where monopoly or oligopoly (heavily concentrated industries) was due to government policies restricting entry. Most of the others made their fortunes in competitive industries, often by improving products or reducing costs (e.g., Sam Walton and Wal-Mart, David Packard of Hewlett-Packard). The percentage of inherited wealth derived from monopoly may be larger (Scherer and Ross 1990).

This argument may be strengthened when we consider some other facts. First, the goal of achieving a monopoly position may itself be socially useful. For example, one important method of achieving monopoly is through patents. However, patents are a case where society has made an explicit decision that the benefits from innovation are sufficiently large so that it is worth granting a monopoly position to inventors. A monopoly may also be achieved through cost-reducing technologies or other productivity increases, and again, these are socially useful.

Second, it is not commonly true that predatory methods are the efficient method of achieving monopoly; that is, one does not generally become a monopolist by driving rivals out of business. Economists generally agree that such behavior is the exception rather than the rule; Shughart (1997, 460–463) summarizes the economics of such behavior and explains why it is generally not profitable. The basic point is that it is difficult for a firm to make money by predatorily lowering price to drive out rivals and then raising price to monopoly levels to regain the earnings lost during the period of predation. This is because once price is raised, other firms can enter and make profits by undercutting the higher price. Examples of successful predation are few; even the famous Standard Oil case, usually considered as the prototype of predation, when carefully analyzed, indicates that Rockefeller did not engage in predation (McGee 1958).

Third, much of what is perceived as monopoly power is in fact due to efficiencies of large size in production rather than an attempt to monopolize, and thus attempts to break up such monopolies might have significant welfare costs. That is, the monopoly provides benefits to consumers as well as to the monopolist, and the elimination of monopoly in such cases will harm, not benefit, consumers.

I do not mean here to defend monopoly. Monopolization exists in the American economy and in other economies as well, and such monopolization does have costs. Moreover, it is possible to become wealthy by becoming a monopolist in circumstances where this would impose costs on consumers. My point, however, is that this is not the primary method of becoming wealthy and that, moreover, noneconomists seem to believe, much more than economists, that monopolization is an important method of gaining wealth.

For example, a recent survey compared economists with noneconomists on several issues (Caplan 2001). It found that noneconomists were much more likely than economists to believe that "business profits are too high," that "top executives are paid too much," and that "oil companies trying to increase their profits" was much more responsible for the "recent [1996] increase in gasoline prices" than was "the normal law of supply and demand," (101) the economists' favored answer. An earlier survey of economists, found that of the respondents 75 percent "generally disagreed" with the proposition that "the fundamental cause of the rise in oil prices of the past three years is the monopoly power of the large oil companies" (Kearl et al. 1979, 30). Alston, Kearl, and Vaughn (1992) find similar results for economists' beliefs about oil price rises in connection with the Iraqi invasion of Kuwait. It is unlikely that general surveys of citizens would give such low weight to monopoly.

We must distinguish monopolization through acquisition of rivals or through predation from monopoly through efficiency. If a firm becomes a monopolist because it is more efficient than others, consumers benefit because they are buying the product at a lower price than they could obtain from others. Similarly, if a firm becomes a monopolist by inventing a new or more efficient process or product, then again, consumers benefit. Only through certain inefficient practices (which are generally thought by economists to be rare) is monopoly harmful.

Productive Activity

The other side of the argument is that things that serve to increase the wealth of individuals are, in general, socially productive. This argument in economics goes back to Adam Smith's famous discussion of the invisible hand. Basically, the argument is that the best way to accumulate wealth is to produce some good or service which others want to buy or to devise a more efficient (cheaper) way of producing some already existing good or service. In either case, this behavior will serve to advance the common good, although of course it is generally undertaken for private gain. That is, individuals undertake such activities not because they want to advance the common good but because they want to increase their own incomes. Nonetheless, economic theory has shown convincingly that (except in the case of fraud or theft and sometimes monopolization) those things that serve to advance one's own welfare will, in the general case, also serve to advance the general level of well-being of other individuals in society. To the extent that this is true, beliefs that the rich are immoral or in some sense undeserving are misguided. Moreover, policies that penalize wealth, to the extent that they are based on false intuitions, are counterproductive.

The argument does not depend on any statement of the moral worth of the rich. Economic efficiency is served and consumers benefit, for example, if resources flow from low-valued uses to high-valued uses. A person who spends effort seeking out bargains (perhaps by seeking out the ignorant and buying goods from them and reselling the acquired items at a profit) may not be a pleasant person. Nonetheless, such behavior is economically useful and does serve to increase the value of resources. Note again that the rich person may not care about advancing general economic well-being and may not even know that he is doing so. This does not matter: to an objective outside observer, the behavior clearly increases the general level of well-being of society. Of course, the rich themselves (or, more likely their children) may not understand the benefits of money-making activities and may, therefore, feel guilty for what were actually productive activities. Schumpeter (1942/1950), a well-known economist of a previous generation, discussed the inability of the bourgeoisie and their children to understand the benefits capitalism provides and argued that this would be one factor leading to his predicted (fortunately, incorrectly) demise of capitalism.

One other example may serve to illustrate the point. Consider speculation—buying some good and keeping it until the price rises. Such behavior is almost universally condemned; the classical Greeks, for example, seemed to have executed persons for engaging in speculation (see Lysias' Speech, "Against the Corn Dealers" in Adams 1905). If the theory advanced here is correct, speculators would have been disliked, for this appears to be exactly the behavior that would have been against the food-sharing ethic which, it has been argued, would have led to survival in hunter-gatherer societies. In such societies, anyone who had food when others were hungry would have been shirking his duty to share and provide insurance for others and, hence, would have been a fit target for condemnation.

Now consider a market economy. Assume, for example, that crops fail somewhat and when the next harvest is reaped, wheat will be in short supply. When the next crop comes in and is harvested, the price will be higher than it would have been without the crop failure. A speculator who learns about this failure can profit by buying now and holding the wheat for future sale at the higher price. The speculator will engage in this behavior to make a profit; if people learn what he is doing, he will perhaps be condemned. Many may erroneously believe that the speculator, rather than the crop failure, has caused the price increase.

Nevertheless, this act of speculation performs a socially useful function. The price will begin to increase now, rather than when the reduced crop comes in, because the speculator is buying wheat and taking it off the current market. But if wheat is scarcer (as it is as soon as the crop failure becomes known), then we want to begin to economize on its use sooner. How do consumers know to economize on the use of a product? They learn this from the price of the product; an immediate increase in the price of wheat will lead consumers to immediately become prudent in their consumption of the product. As a result, in the future, when the poor crop is harvested, the shortage will be less severe since more stored wheat will exist because of the earlier price increase and the resultant reduction in consumption. Thus, the speculator who forces the price increase sooner than would otherwise have been the case will actually perform the socially useful function of causing greater efficiency in the use of the now relatively scarce commodity. That he does so for essentially selfish reasons is irrelevant; the fact remains that a useful function is being performed. Moreover, economists have known this for some time, yet speculation is still widely viewed as immoral.

Saving and Investment

Most wealthy persons in today's society achieved their wealth through the mundane methods of frugality and saving (Stanley and Danko 1998). Such behavior is efficient. Saving creates a pool capital that firms use for making investments in capital goods, which, in turn, increase the productivity of workers and the general wealth of society. Such behavior is socially useful and productive. However, as argued above, we may not have clear intuitive understanding of the productive value and benefits from investment and capital, and so may misunderstand or underestimate the social benefits of such saving.

Hierarchy

We can see the role of envy if we analyze hierarchies and the way that people think about them. Hierarchies (also called dominance hierarchies or pecking orders) are virtually universal in social primates and other group-living nonhuman animals.[1] Hierarchies are also universal in human societies. But while some human hierarchies are similar to animal hierarchies, some are not. Humans are often confused about the different types of hierarchies, and students of human behavior and evolution also suffer from this confusion. Our understanding of human society will improve if we distinguish two uses of hierarchy.

The basic distinction is this: animal hierarchies are used for allocation of a relatively fixed amount of resources, including sometimes access to females. Their use is essentially zero sum. I call these allocation, consumption, or dominance hierarchies. Some human hierarchies are of this sort. Moreover, this form of hierarchical organization is evolutionarily old and predates humanness. Such hierarchies are common to many other primate species (e.g., de Waal 1996, chapter 3), and, therefore, attitudes and behaviors with respect to these structures would have evolved in various ways. Humans have created a new use for hierarchies, however. They use hierarchies for productive purposes. That is, many productive human activities—for example, all the activities that occur in business firms, in governments, and in universities—are organized in hierarchies. In contemporary society virtually all productive activity is at least in part hierarchically organized. Such hierarchies seem to be relatively new for humans (i.e., new relative to the length of time of human evolution). I call these productive hierarchies.

Humans often confuse the two types of hierarchies. Moreover, productive hierarchies do have some features in common with dominance or consumption hierarchies and were probably developed from them. Therefore, it is not surprising that humans (including students of human evolution and behavior) often confuse the two. But there are also fundamental differences and sometimes errors in understanding or in policy are motivated by this confusion. That is, economic and social policies—often very important ones—dealing even with the overall structure of society—are sometimes motivated by confusion between production and consumption hierarchies. Moreover, scientific analysis of human behavior would be improved if there were a better understanding of the difference between types of hierarchies.

Consumption Hierarchies

Although hierarchies arise from individual behavior, they also provide a social benefit. The benefit of the hierarchies studied by students of evolution is generally to allocate resources with reduced amounts of intragroup violence. The resources allocated include food, territories for production of food, and (for male hierarchies) access to females. Most allocation that occurs is zero sum. That is, there is a fixed amount of resources (food, land, females) to allocate, and the hierarchy determines the way these resources are allocated, with more dominant members generally having preferred access. Geary (1998) indicates that male dominance hierarchies are more commonly used for allocation of females, while dominance hierarchies among females are often used for allocation of food.

Though these hierarchies do not lead to increased production, they do serve a useful purpose in that they reduce conflict within the group, which would otherwise lead to reduced levels of resources. Moreover, each individual—even relatively low-ranked members of the hierarchy—may benefit from its existence, or at least not be in a position to change the situation. This is because all individuals obtain the benefits of group living (generally protection from predators, sometimes including members of the same species) and have at least some chance of ultimately rising in the hierarchy. Thus, there is no need to posit any form of group selection to explain hierarchies. They can exist and maintain themselves purely as a result of normal individual selection. This also implies that, though hierarchies may have a useful social function, this function arises as a result of individual maximizing behavior as individuals compete for scarce resources, not as the result of any plan or goal. It is interesting that discussions of the role of dominance hierarchies by biologists and anthropologists are always in terms of allocation.

Among nonhuman primates, the nature of hierarchies can vary. Some are rigid, with dominant individuals demanding extreme obedience. For example, the rhesus monkey hierarchy is very structured, and dominants eat all they want before subordinates eat at all. Others are much more open and less structured. Among chimpanzees, subordinates often successfully beg for food from dominants. De Waal (1996, chapter 3) has a discussion of hierarchies in primate species, including some discussion of human hierarchies. He also indicates that an additional function of alpha (high-ranking) males in many primate groups is mediation of disputes among lower ranked individuals. There is political competition for dominance in primate societies. This often involves coalition formation, and much of de Waal's research has been in documenting this competition. Among many species, including chimpanzees, our nearest relatives, this competition includes forming alliances and other political manipulations.

De Waal also indicates (1996, 127) that the amount of egalitarianism within a species is a result of the interplay of two forces: the value of cooperation to the groups, and the ease of leaving the group. As the amount of either of these factors increases, the species becomes more egalitarian and less hierarchical. This is because increased value of cooperation means that if individuals leave the group, the group loses more, so dominant individuals must refrain from misusing their power. As ease of leaving the group increases, again dominants are limited in their ability to use their power. Increased value of cooperation is generally due to increased value of defense—from predators, including often conspecifics. De Waal's argument may be expressed as an economic argument. The value of cooperation is essentially the demand for additional group members. As demand increases, price (measured by the amount of autonomy given to members) increases. Ease of leaving is the inverse of supply. An increase in ease of leaving the group is equivalent to a reduction in supply, and a reduction in supply leads to an increase in price. De Waal provides no examples of any productive use of hierarchies among nonhuman primates other than the suppression of internal conflict. Thus, the hierarchies he describes are examples of what I call consumption hierarchies.

In hierarchical species, both males and females form hierarchies. And, hierarchies are stronger and better defined among the more strongly bonded gender. This in turn depends on which gender migrates and which remains in place in a given species (Geary 1998). Among ancestral humans and our closest relatives, males remain in place and young females migrate; these species are said to be patrilocal. This pattern has important implications for several aspects of human behavior dealing with coalition formation among males, as discussed in chapter 2. For the purposes of this chapter, what is relevant is that among humans, male hierarchies are traditionally more important and ranks are more clearly defined than in female hierarchies. In most other primates, males migrate and females remain in place. In these species, females often form coalitions with close relatives and compete with other kin-based female coalitions for valued resources, such as control of fruit trees. On the other hand, in chimpanzees, which are also patrilocal, male bonds seem stronger than female bonds. Female chimpanzees spend 50 percent of their time with only their dependent offspring; for males, only 18 percent of time is spent alone (Pusey 2001).

Dominance hierarchies are also universal among humans, at least as group size becomes larger than the band. Geary (1998) discusses these hierarchies at several places. The hierarchies are more significant and well defined for males, as is consistent with the primate pattern for a partrilocal species. Most human societies have been polygynous, and more dominant men, who have gained control over socially valuable resources, use these resources to obtain more wives and more offspring (in biological terms, to become more fit, or to have more reproductive success) than others. In traditional hunter-gatherer societies, where most human experience has occurred, the struggle for dominance is ongoing, and homicide as part of this struggle is fairly common and is much more widespread than in modern societies. Edgerton (1992, chapter 4) describes in some detail several societies where dominance hierarchies caused much human misery among the lower ranked members, including the Kwakiutl, the Aztecs, the Zulus, the Asante, and, more recently, Stalin in Russia and the Khmer Rouge in Cambodia.

Within human dominance hierarchies, risk taking to achieve higher rank is worthwhile, particularly for young males because it may be necessary to rise to a certain level in the hierarchy in order to reproduce at all. Any male who does not rise to this level will not reproduce and will have left no genes in the population. Males who do not take the risks required to rise in the hierarchy will be genetically equivalent to males who take risks and fail; both will leave no descendants. Only those males who took risks and succeeded have become our ancestors. This argument is exactly opposed to a group selection argument. The group would benefit if all males lived and served as defenders of the group, because this would increase the ability of the group to defend itself or to predate against other groups. Nonetheless, males will take excessive (from the viewpoint of the group) risks for private reasons having to do with their own fitness. In chapter 6, we will see how this tension leads groups to trying to regulate the risk-seeking behavior of young males. Daly and Wilson (1988) show that even in contemporary societies, a large fraction of homicides are due to dominance struggles between young males. Other forms of risk-taking behavior by young males—fast and dangerous driving, excessive use of alcohol or drugs, perhaps gambling—are forms of competition for higher positions in dominance hierarchies, where one goal is to show courage.

Even when hunter-gatherers engage in combat, the form is generally not based on a hierarchical military, such as we are accustomed to. Rather, the typical form of combat is raiding, where a party of men from one village will ambush one man from another (Wrangham and Peterson 1996). Warfare does not seem to be organized, and warriors are not arrayed in a hierarchical system, even though successful warriors (usually, those who have killed an enemy) will rise in position within the dominance hierarchy and will have access to more resources and often more females in their own society. In his discussion of primitive hunting, Stanford (1999) does not mention any hierarchy in the process. Ridley (1996) describes trade and exchange among hunter-gatherers and explains the gains from trade but has no discussion of hierarchy used in production. I discussed Ridley's arguments in chapter 1, when I considered division of labor in primitive societies.

But while hierarchies are universal among humans, opposition to these hierarchies is also quite common. Boehm has written extensively about this resistance in hunter-gatherer societies (e.g., Boehm 1993; 1997a and 1997b;

1999). A strong statement of Boehm's hypothesis is: "I believe that as of 40,000 years ago, with the advent of anatomically modern humans who continued to live in small groups and had not yet domesticated plants and animals, it is very likely that all human societies practiced egalitarian behavior and that most of the time they did so very successfully" (Boehm 1993, 236). Boehm shows that hunter-gatherers used methods ranging from ridicule to exile to homicide to restrain the ambitions of would-be dominants.[2] He argues that there is a reverse dominance hierarchy, in which potential dominant individuals are controlled by their putative subordinates. Knauft (1991, 395) makes a similar point: "there tends to be active and assiduous devaluation of adult male status differentiation and minimization or denial of those asymmetries that exist."

Boehm is uncertain as to whether the tendency to avoid hierarchies itself led to genetic changes. At one point (1997b, S115), he indicates that it may be that the tendency of humans to avoid hierarchies "did not necessarily involve any major changes in the human genotype, for basic tendencies to dominate and submit and tendencies to form coalitions may have been all that was needed, along with a shared will to avoid dominion and the political intelligence to cooperatively head it off." Elsewhere (1997b, S117) he suggests that "the genetic basis for this ambivalence was structured by the levels of selection that prevailed prehistorically, at a time when the egalitarian syndrome reigned universally. . . ." More recently, he seems to have decided that there was a genetic effect (Boehm 1999, chap. 9). However, we need not resolve this issue. It is enough to note a tendency among humans to resist being low ranked members of dominance hierarchies. I discuss these issues more fully in chapter 5, where I discuss issues related to political power.

Production Hierarchies

As discussed in chapter 1, in the EEA there was probably little specialization and division of labor except by gender within a family or bonded pair. Not until societies became sedentary did complex patterns of exchange and specialization begin. Of course, in today's economy massive division of labor and specialization are endemic and responsible for much of our wealth.

If there is division of labor, then no individual or family is self-sufficient. Individuals then need a method to obtain those goods that they do not produce for themselves. This requires some economic coordination. There are two such methods of coordination: through the market and through command and control, or hierarchies. Economists have had some difficulty understanding hierarchical methods of production because economics generally study allocation through markets and firms and other examples of hierarchies that replace the market for some transactions and exchanges.

An early paper analyzing this issue was Coase (1937). More recent sources include Klein et al. (1978) and Williamson (1983) (see also Rubin (1990) for a general analysis). The point is that production can be coordinated by markets or within firms. Some activities are better coordinated through the market and others through hierarchy. Which method is preferable for any given activity depends on transaction costs. When the possibility of cheating through expropriating fixed investments becomes too large, hierarchy replaces the market. An example from Klein et al. (1978) is that General Motors ultimately purchased Fisher Body because the possibility of Fisher holding up GM became too great.

The level of costs associated with hierarchy itself depends in part on evolved human preferences for alternative working environments; Rubin and Somanathan (1998) discuss this point. The observed organization of production will be that which is cost minimizing for firms, but the costs include payments that must be made to workers for working in a manner viewed as less desirable because of evolved preferences for sociality in the workplace. This may be why predictions of increased telecommuting seem to have been overdrawn. Workers may not like working alone at home, and the increased payment demanded for such work may often outweigh any increases in productivity. But the clear implication is that some hierarchically organized entities are productive and that hierarchies do not exist only for the purpose of allocation.

In other words, humans have adapted the dominance hierarchy to new uses related to division of labor. Among humans, hierarchies are used to perform tasks. Superiors induce subordinates to undertake coordinated activities whose benefits are shared in some way throughout the hierarchy. There is division of labor within the hierarchy. Sometimes hierarchies are used to perform tasks related to aggression, as in the use of armies. Hierarchic armies are more productive (of violence) than are the unorganized raids that Wrangham and Peterson (1996) discuss. Sometimes hierarchies are used for actual production of goods and services, such as pins in a pin factory or education and research in a university. This use of hierarchy appears to be a little appreciated human innovation and is probably relatively recent in evolutionary terms. During much of human evolutionary time, hierarchies would have been mainly consumption hierarchies. With relatively little specialization, there would have been relatively little gain from productive hierarchies. Therefore, the main experience in the EEA with hierarchies would have been with dominance or consumption hierarchies, the sort that Boehm and Knauft indicate that subordinates resist.

The key distinction relevant for understanding productive hierarchies seems to be between what are called simple and complex hunter-gatherers (Knauft 1991; Kelly 1995.) Simple hunter-gatherers are mobile, egalitarian, live in small settlements, and the only occupational specialization is by age (Kelly 1995, table 8-1, 294.) Complex hunter-gatherers differ in many dimensions; in particular, occupational specialization is common. Thus, this distinction (which Kelly attributes to changing from a mobile to a sedentary life style) is the border between hierarchies used for dominance and for production. By the time of large agricultural societies, which arose out of complex hunter-gatherer societies, division of labor was universal and important among humans. Production hierarchies probably became important with the beginning of sedentary societies. Bruce Smith (1995) indicates that settled societies had very complex physical structures, which probably would have been impossible to create without division of labor (see also Kelly 1995). Agriculture originated in such societies, and division of labor was clearly important in agricultural societies. Diamond (1998) argues that agriculture created the possibility of food storage, which led to the feasibility of kings and other dominants. Kelly also shows that sedentary societies had complex dominance relationships and also food storage. The causation may be in some other direction; perhaps increased productivity of farming or sedentary living, rather than simply the possibility of food storage, led to the effects Diamond identifies. Or hierarchical methods of production, such as coordinated efforts at irrigation, led to the increased availability of food.

Productive hierarchies probably developed from allocation hierarchies. A plausible scenario is that some dominants began using their power to induce subordinates to engage in coordinated productive activities, which may have included organized predation. (Predation, while negative sum for the totality, can be productive for the victors.) From here, two potential paths are possible. One path would have been to modify the allocation hierarchy to make it more productive. That is, the order of dominance of members of the hierarchy might have been rearranged to increase productivity. A second step would have been to create entirely new hierarchies for productive purposes. It is clear that at some point this step was taken, for today productive hierarchies are routinely founded. This is an important difference between allocative and productive hierarchies. The former are inherent within a social species, while the latter are generally consciously and purposefully created to serve a specific purpose.

Once such productive hierarchies developed, two factors did develop together. First, chiefs and then kings and other rulers became despotic leaders of hierarchies. Betzig (1986 and elsewhere) has shown that such dominant individuals were able to engross for themselves very large numbers of women, forcing others in society to remain unmarried. This demonstrates the wisdom of those hunter-gatherers who repressed tendencies toward dominance. Second, the general and widespread use of hierarchies for production began with the advent of sedentary and then agricultural societies. Indeed, this extra productivity may have led to the beginning of states and other governance structures with centralized power. One hypothesis is that of Wittfogel (1957), who argues that it was the increased productivity of irrigation that caused such kingships. We need not accept his hypothesis completely to accept that the possibility of increased productivity through hierarchy was important in generating political institutions and dominance. Richerson and Boyd (1999) discuss the productive role of hierarchies in the governance of complex societies. Sober and Wilson (1998) also indicate that hierarchies may be productive. Until relatively recently, most increased productivity was translated into increased population, but in more recent times (at least in the West), it has been used for increased consumption within a fixed or even shrinking population.

Both production and consumption hierarchies have many features in common. In particular, those who are higher up—more dominant—in the hierarchy receive more resources in both types of hierarchies. Additionally, superiors are able to give commands to subordinates. This is because productive hierarchies derived from the evolutionarily older dominance hierarchies and retained many features of the older structures. Thus, any innate or evolved preferences (or antipathies) regarding dominance hierarchies may well be applied to productive hierarchies as well. As we see later, scholars studying human evolution sometimes confuse the two types of hierarchies, and ordinary humans may do so as well. Moreover, even in productive hierarchies, dominants may appropriate a disproportionate share of the increased output, so some resentment of some productive hierarchies may be justified.

For individuals to join a hierarchy voluntarily, there must be either economic or fitness benefits, or both. In modern societies and in a free economy, no one will voluntarily join a hierarchy unless he is compensated sufficiently, so economic benefits (which can be translated into fitness benefits) are a requirement. That is, the existence of competitive labor markets means exactly that no one will join a hierarchy unless the benefits of this hierarchy are greater than any alternatives. This requirement is also the participation constraint in economic models of principal-agent interactions (Arrow 1986). In the past, individuals may have been coerced into joining hierarchies, mainly governmental, so that such benefits would not have been assured.

There are circumstances under which lower ranked subordinates within a hierarchy could do less well in terms of fitness from being members of even a productive hierarchy. This could occur even if lower ranked members received compensation in excess of their alternative earnings. This would be the case if the higher earnings of dominant members of the hierarchy were sufficiently great so that they could accumulate additional wives. It would also be necessary for the hierarchy to be sufficiently large relative to the society in which it existed so superiors accumulating additional females would have an effect on the lower ranked members of the hierarchy itself. That is, if one small productive hierarchy were embedded in a large polygynous society, lower ranked members of the hierarchy could acquire wives even if their superiors were polygynous. But in the EEA, societies were small, and any hierarchy would have been large enough for this condition to be violated. Then the lower ranked members could have larger incomes but reduced fitness; this is an example of a conflict between the economist's notion of utility maximization and the biologist's notion of fitness maximization. In this circumstance, lower ranked individuals would have resisted even productive hierarchies. (An exception could have been hierarchies for aggression: conquest of a neighboring tribe or band would generally have led to additional women for the victors.)

In modern western civilization, this conflict does occur, because in these societies there is a doctrine of socially enforced monogamy, and superiors, no matter how wealthy, cannot acquire a large number of wives (Alexander 1987; MacDonald 1995). Some have stressed the value of such monogamy for social stability (e.g., Alexander 1987), and others have argued that increasing value of subordinates generated through division of labor caused elites to be willing to forgo the fitness benefits of polygyny (Betzig 1995). However, so far as I know, no one has pointed out the relation between productive hierarchies and monogamy. The point here is that, even if hierarchies lead to widely uneven incomes, all members can still benefit. This will occur if the hierarchy leads to increased productivity and increased earnings by subordinates and if socially imposed monogamy prevents dominants from using their increased wealth to accumulate additional mates at the expense of subordinates.

Though hierarchies today are generally beneficial for all their members, fear or dislike of steep hierarchies may rest on a biological or evolutionary base. This notion is consistent with Boehm's arguments and with tastes that evolved in a world where all hierarchies were dominance hierarchies. If so, then people may resent even productive hierarchies and the associated income distribution, even if they actually benefited in terms of fitness and of wealth. I address this point more carefully below.

Government Hierarchies

Today, some productive hierarchies are associated with government. Governmental hierarchies share two characteristics with classic dominance hierarchies but not with other productive hierarchies. First, each society has only one government hierarchy.[3] Second, everyone in a society must be subject to the government hierarchy. Neither feature applies to other production hierarchies. There are many productive hierarchies within a society, and a large organization such as a modern firm or university will itself contain literally hundreds of subhierarchies. Moreover, membership in most hierarchies is voluntary. While it would be difficult for me to leave the United States, it would be relatively easy to leave my department or university. Universal hierarchies with limited exit characterize government hierarchies; numerous hierarchies with easy entry and exit characterize other productive hierarchies. Anthropologists pay relatively more attention to government hierarchies than to other productive hierarchies. This has added to the confusion between different types of hierarchies, discussed later.

Fears of excessive dominance by government hierarchies, of the sort that Boehm emphasizes, may be justified, because exit from such hierarchies is difficult. For example, many governments (including that of the United States until recently) coerce individuals into joining a military hierarchy, even if it is not in their perceived self-interest, through the military draft, and governments routinely collect taxes and impose regulations that may be unpopular. Of course, at various times some individuals have found emigration a preferred strategy to the draft or to other government policies, but this choice is generally a minority decision. Such emigration is costly, but it does limit the possibility of coercion. Particularly oppressive societies, such as the former communist societies of Russia and Eastern Europe, limited the possibility of emigration to increase possibilities of coercion. More generally, a form of bottom up control in democracies is aimed exactly at reducing the coercive power of the government. Of course, this is a weaker constraint than the possibility of easy exit, which limits the power of nongovernmental hierarchies. Nonetheless, in modern democracies government control of individuals is significantly less than in earlier societies, except for the hunter-gatherers Boehm described. (For a general argument that industrial society is freer and less repressive than society in the past, see Maryanski and Turner 1992).

Confusion between Types of Hierarchies

Students of evolution have done well to point out that human hierarchies are similar to pecking orders and other dominance hierarchies in animals, and I do not want to detract from that work. Nonetheless, in many cases the analogy has been overdrawn. I have not done a complete survey of the literature, but I have read widely and have found no cases in which the distinction is made between allocative and productive hierarchies. On the contrary, I have seen examples where arguments appropriate to dominance hierarchies have been inappropriately applied to contemporary production hierarchies, as exist in a modern economy. Some examples follow, but I want to stress that these should not be read as being critical of the works cited. The point is a subtle one, and the purpose of this analysis is to point out the distinction.

Boehm comes close to making the point. He indicates that hierarchies can be useful for "protection from aggression or economic risk" and "economic redistribution" (1993, 246.) He does not make the point, however, that such hierarchies can be useful for increased production as well as redistribution. Geary has a nice discussion: "For the most part, however, boys and men more typically compete physically or through the attainment of socially prized resources, such as money or the heads of one's competitors. . . ." (262). This exactly makes my point: seeking the heads of one's competitors is clearly a negative sum endeavor, while seeking money is associated with increased productivity and a positive sum effort.

Singer (2000, 38) commits the same fallacy: "For example, if we live in a society with a hierarchy based on a hereditary aristocracy, and we abolish the hereditary aristocracy, as the French and American revolutionaries did, we are likely to find that a new hierarchy emerges based on something else, perhaps military power or wealth." A hereditary aristocracy is an allocation hierarchy; a military hierarchy is based on predation; a hierarchy based on wealth is a productive hierarchy. To treat these identically is to miss the productivity of modern hierarchies and to focus on the distributive aspects of hierarchy. I have already mentioned Diamond's (1998) discussion of the beginning of agriculture, in which he analyzes stockpiling of food and expropriation of food by a political elite but does not consider the possibility of increased productivity. Knauft (1991, 397) discusses similarities between hierarchies in great apes, which are clearly dominance hierarchies and in middle range human societies. (Knauft believes that these similarities are due to analogy rather than homology, because he believes that for much of human evolutionary history there were no hierarchies.)

If professional students of human behavior and evolution have confused consumption and production hierarchies, it would not be surprising if other humans made the same error. The same factors that Boehm identifies as leading humans to dislike dominance hierarchies can lead to a dislike of productive hierarchies, even though these may lead to sufficiently increased output so that all members can benefit. In particular, humans may be overly hostile to productive hierarchies and may as a result choose policies that actually make them worse off—from both an economic (utility or wealth maximizing) and biological (fitness or reproductive success maximizing) perspective.

The most powerful and most tragic example is Marxism, or communism. I claim that Marx's opposition to capitalism and the acceptance of Marxism by many individuals (including many intellectuals) was based on confusion between productive and consumption hierarchies of exactly the sort addressed in this chapter. Marx did realize that capitalism was a highly productive system, but his analysis seemed to confuse characteristics of productive and allocation hierarchies. The *Communist Manifesto* (Marx 1888), the basic policy doctrine of communism, reads like a discussion of dominance hierarchies.[4] "In the earlier epochs of history we find almost everywhere a complicated arrangement of history into various orders, a manifold gradation of social rank" (1). Also: "Modern industry has converted the little workshop or the patriarchal master into the great factory of the industrial capitalist. Masses of laborers, crowded into the factory, are organised like soldiers. As privates in the industrial army they are placed under the command of a perfect hierarchy of officers and sergeants" (6). And: "The history of all past society has consisted in the development of class antagonisms, antagonisms that assumed different forms in different epochs. But whatever form they may have taken, one fact is common to all past ages, viz., the exploitation of one part of society by another" (15). Marx even considered the reproductive effects of his schema in a rather bizarre passage: "Our bourgeois, not content with having the wives and daughters of their proletarians at their disposal, not to speak of common prostitutes, take the greatest pleasure in seducing each other's wives" (14).

Marx's arguments would be appropriate if the hierarchies he described were dominance hierarchies, with the dominant individuals viewed by Marx as the bourgeoisie. In particular, the discussion of factory workers as being "placed under the command of a perfect hierarchy" misses the point that workers voluntarily chose to work in factories, because the additional compensation over any alternative was sufficient to compensate for being "under the command" of the managers (Marx, 1888, 6). This additional compensation was available, of course, because workers in factories were more productive than in alternative workplaces, and part of this extra productivity went into compensation. Owners were and are forced to return this additional compensation to workers by forces of competition.

If the claim that the bourgeoisie did have "the wives and daughters of their proletarians at their disposal" were true to a significant extent, then the Marxist anticapitalist position would have been biologically justified, even if it led to reduced output and reduced incomes. That is, this could describe a situation where wealth maximization and fitness maximization would have led to different outcomes. But of course capitalism grew up in a society with socially imposed monogamy; one argument of this chapter is that it was only in such a society that capitalism could have come into being.

Therefore, the Marxist anticapitalist position would have increased neither wealth nor fitness of workers. Moreover, the appeal of Marxist belief—which persists in some circles even today, when the dismal implications of a communist society should be clear—was based on the human opposition to dominance hierarchies, inappropriately applied to productive hierarchies. For example, many anthropologists, who might be expected to understand the distinction, use a basically Marxist mode of analysis and may thus be unable to ascertain the productive role of hierarchies. For example, Roseberry (1988) and Kelly (1995, 29-33) discuss the role of Marxist arguments in anthropology.

In terms relevant to anthropologists and others, we may say that Marxism is subject to the same criticisms as naïve group selectionism. Marxism assumes that individuals maximize their class interests. (Neo-Marxists have added race and gender.) A Marxist argument that the capitalist class would collude to exploit workers (or a neo-Marxist argument that men collude exploit women or whites to exploit blacks) is subject to exactly the same free-riding forces that make classical group selection unfeasible. Any one capitalist would have benefited by cheating on the collusive agreement and, for example, increasing wages offered to attract more productive workers. It is therefore difficult to understand why those who often reject group selectionist arguments nonetheless accept Marxism. For example, Kelly's statement (1995, 49) that "cultural ecology's . . . adherence to an outdated group-selectionist perspective caused many anthropologists to turn to alternative theoretical frameworks, such as Marxism" is odd because Marxism is itself a group selection theory and suffers from the weaknesses of such a theory.

Another error made by Marx and accepted by many others may be based on evolved patterns of thinking. Recall that there was little capital in the EEA, and as a result I argued that we may not have reliable intuitions about the productivity of capital. This may explain the Marxian labor theory of value—the theory that all value is created by labor. This theory is clearly incorrect, but it may be intuitively appealing for evolved reasons. This may also explain why many religions forbid the payment of interest. Interest is a payment for the use of capital, and if one does not understand the productivity of capital, it is impossible then to understand the value of interest.

Redistributive Policies

The issue of the nature of hierarchies is also relevant for understanding preferences about income redistribution. There are two types of policies aimed at income redistribution and distinguishing between them is important. Some policies are designed to increase the incomes of the poor, as discussed in chapter 3. Preferences for these policies would have evolved in the small communities that were standard in the EEA, and they can be understood in terms of reciprocal altruism, kin selection, or efficient altruism. These policies are based on feelings of sympathy for the poor or unlucky. Other policies are aimed at reducing the incomes of the wealthy. These must be understood in terms of Boehm's arguments regarding dominance: These policies are remnants of the desire of our ancestors to reduce the power of dominants. They can be understood in terms of feelings of envy or jealousy.

In many cases, the same policies will satisfy both goals, as when taxes on the wealthy are used to generate revenue to supplement incomes of the poor. But in some cases the policies work at cross-purposes, as when tax rates on the wealthy are so high as to actually discourage effort and reduce the total amount of income available. Roe (1998) discusses redistributive policies that have the effect of actually reducing incomes and indicates that the fall in incomes in Argentina in the twentieth century may be due to such counterproductive policies based on envy. Even Boehm (1997b, S117) confuses the two types of policies, when he discusses "social ideologies that are altruistic . . . that range from hunter-gatherer codes of cooperation and helpfulness, to modern welfare-state ideologies, to the idealistic communist blueprints that have proven so very difficult to execute."

Policies aimed at providing benefits for the poor or unfortunate may be understood as a form of altruism, and these policies can achieve their goals in current society. But policies aimed at harming the wealthy are generally counterproductive in a monogamous free society where hierarchies are productive and membership is voluntary and where wealth creation is generally due to efficient activities. Such policies reduce the wealth or income of all members of society, and they also reduce fitness. These policies are based on confusing the different types of hierarchies and on inherited preferences that are now harmful. It is important to understand this basis for policy to help avoid similar errors in the future.

Among those concerned with welfare of the poor, there are two possible positions. It is possible to be concerned with the level of income—how much money do the poor have to spend? It is also possible to be concerned with the amount of equality—how much income does the bottom fifth have as compared with the top fifth? Rawls and Marx prefer equality. Singer (2000, 45) makes an egalitarian argument: "Encouraging the pursuit of self-interest through the market economy may have contributed to a high average level of prosperity, but a the same time the gap between rich and poor has widened, and support for the poor has been cut back."

If we measure the fitness of the poor or their consumption ability, then absolute incomes, rather than relative incomes, would matter. If a poor person in the United States has more resources than a middle-class person in Africa, then the American is in some sense better off. In this sense, understanding why some would be in favor of equality rather than levels of income is difficult. Additionally, if policies aimed at increasing equality lead, for example, to lower economic growth, they can actually lead to more poverty. In fact, evolutionary reasons explain why equality might appear important. First, in a world of polygyny, inequality may matter more than level of income. This is because if polygyny is allowed, the wealthier may engross excess females, leaving the poor with none. Second, in a zero-sum world inequality is the same as poverty for the poor. That is, in a zero-sum world, if some are wealthy, this must be at the expense of the poor. Thus, we can understand how tastes for equality may have evolved among altruists.

But neither of these conditions exists now. We no longer have polygyny, and we no longer live in a zero-sum world. Therefore, in today's world, those interested in helping the poor should focus on policies that increase the incomes of the poor and ignore any effect of these policies on the distribution of income. For example, free importation of goods reduces the prices paid for goods, and so benefits all consumers, rich and poor. This policy would do much to improve the incomes of the poor, and altruists should favor this more than many directly redistributive policies. Similarly, production for export in low-wage countries is the surest way to increase incomes of poor workers in those countries, and those who oppose such production because of the low wages earned by workers are actually proposing policies that harm the people they purport to benefit.

A similar point applies to opponents of globalization who advocate policies that will increase levels of poverty in the guise of helping the poor. (This point is made in any economics text that discusses international trade. For popular discussions, see Krugman 2001, and Kristof and WuDunn 2000.) Similarly, much opposition to various technologies may serve to harm poor people in developing countries, while providing noneconomic benefits to residents of richer countries. This applies to the use of DDT for malaria control and also to the use of genetically modified foods (United Nations 2001). Choices made in developed countries to restrict technologies that would benefit residents of less developed countries may be influenced by the fact that the potential beneficiaries are of different ethnic backgrounds from the decision makers. Therefore, there may be implicit discounting of the gains or losses to the poorer persons. For example, bans on DDT may lead to one million deaths per year, mainly of children in the tropics (United Nations 2001, 69). Citizens of developed countries in favor of such bans might be less willing to tolerate these deaths if they were of more closely related individuals.

Summary

If anyone accumulated wealth in the EEA, it was probably the result of shirking and failing in a responsibility to share. As a result, humans may have developed tastes for disliking the wealthy or tastes including envy. Such tastes seem very common among humans. In today's world, most wealth accumulation is through productive activity, and so envy and tastes for envy are harmful and counterproductive.

In nonhuman species and in hunter-gatherer societies, hierarchies are generally used for allocation of scarce goods, including often access to females. Humans in hunter-gatherer societies resisted hierarchies, so much that Boehm referred to a reverse dominance hierarchy. More recently, humans have taken the evolved form of hierarchy and adapted it to new uses. In particular, among humans, hierarchies are used for productive as well as allocative purposes. The widespread use of hierarchies for productive purposes likely came about at the time of the transition from mobile to sedentary hunter-gatherer societies, because there was probably relatively little specialization in earlier societies.

The socially imposed monogamy in the modern western world does not allow wealthier individuals to use increased wealth to engross excessive numbers of females. Moreover, individuals must be compensated if they are to take subordinate roles in hierarchies. Therefore, in modern societies, productive hierarchies are useful and beneficial for all concerned. But productive and dominance hierarchies have many features in common. In both, higher ranked individuals receive more resources, and higher ranked individuals can issue commands to lower ranked members. Therefore, humans (both normal humans and students of human behavior) often confuse the two uses. For example, the *Communist Manifesto,* a major policy document, clearly suffered from this confusion. For both analytic and policy purposes, the difference must be understood. It would also be useful to understand the point in time when humans began using hierarchies for production and to analyze ways in which humans converted the mechanism of the dominance hierarchy into the productive hierarchy. Since the issue has not been examined in this framework, the question has not to my knowledge been addressed in the literature.

Several issues are relevant for understanding ways of increasing incomes of the poor. Programs aimed at increasing incomes of the poor are useful, but programs aimed at reducing the incomes of the wealthy are not. Similarly, increasing the incomes of the poor is a desirable policy, but increasing equality is not. We may have preferences for equality based on zero-sum thinking, but in a world of monogamy such programs are neither fitness nor wealth increasing for anyone.

CHAPTER 5

Political Power

Contemporary humans seem to want freedom. In the United States, the American Revolution was at least in part about freedom, and freedom and liberty are perennial terms used in political discussions. At least since the defeat of the Nazis, Western Europe has also had relatively free political systems. The Communist countries always characterized themselves as democracies or people's democracies, even though they were undemocratic and the people had relatively little freedom. When it became possible, the citizens of these countries were willing to accept some risks to achieve freedom. Parts of Africa and Asia are relatively less free, and continual struggles for freedom in those places persist.

But throughout most of human history people have had relatively little freedom. History books are full of stories and accounts of kings and dictators, and much of history is an account of struggles between various kingdoms, with relatively little attention paid (by either historians or by the kings and dictators) to the welfare of the people. Many preliterate societies have also been ruled by what must be called evil dictators (Edgerton 1992.) The desire for freedom might appear to be something new—something that was invented in the seventeenth century.

Actually, a desire for freedom is an evolutionarily very old characteristic of humans, and throughout most of human existence, most humans (or, at least, most males) have been quite free. The perspective obtained from studying *history* is biased, because throughout most of the period of writing, humans have lived in an unnatural state, a state of reduced freedom.[1] For most of prehistory—which represents most of the time humans have existed as humans—people were free. During one lengthy period (about 10,000 years) circumstances were such that most humans enjoyed only small amounts of freedom, but this was an aberration.

Of course, humans, in addition to wanting to be free themselves, also want to be dominant. That is, individuals prefer to be dominants and reduce the freedom of others.[2] Sometimes subordinates can resist this desire for power by dominants, and we have freedom. At other times, circumstances are such that subordinates cannot resist this power, and we have monarchy or dictatorship. This chapter illuminates this struggle from the perspective of evolved tastes for freedom and for dominance. We start with the seeking of political power.

By now, it should be clear that political power seeking has generally been a man's game. (Hrdy 1999, 252: "I know of no evidence that any matriarchal society ever existed"; see also Eller 2000.) Low (2000) points out that, although the status of women's rights varies cross culturally, 70 percent of societies have only male political leaders. There are several reasons for this. Political success is more valuable to men then to women, so men have a higher demand for such activity. Additionally, political activity is less costly for men, so supply is greater. Since both demand and supply factors work together, men engage in more political activity.

From the demand side, the value to men of additional power, in the form of extra wives or mates, has been important, and so those males who sought and obtained such power left more descendants than did those who did not seek power. The biological basis for power seeking itself is that this behavior was associated with greater fitness in evolutionary times for humans and our ancestors. Because biological factors limit the number of potential offspring for a woman, this source of value did not exist for women. Indeed, males seeking political power is something we have in common with chimpanzees (de Waal 1996). Additional power brings three types of fitness benefits. First, power leads directly to acquisition of more females from within the group. Second, power leads to acquisition of resources such as food that can then be translated into increased fitness through acquisition of additional females and also through increased survival of the male himself and of his mate and offspring. Finally, an additional function of politics has been to organize to more effectively predate on neighbors or to defend against such predation, and these forms of predation have always been a male's game as well. This is because a major benefit of such predation has been the ability to capture females from the losers.

Male dominance in political power can also be explained through cost and supply. As discussed in chapter 2, human and proto-human societies have been patrilocal, so males have had relatives nearby. Coalition formation, the essence of politics, has been easier (less costly) for males, because they already have ready-made partners in the form of fathers, brothers, uncles, and cousins. Because of this genetic relatedness, cheating in the prisoner's dilemma associated with coalition formation is less likely. Since women migrate, a female usually will not have close genetic relatives to associate with, and so coalition formation is more difficult.

As a result of these factors, men view politics as being more important than do women. In the survey of "Human Values and Beliefs," (Inglehart, Basañex, and Moreno 1998) in all countries except one,[3] a higher percentage of men than of women say that politics is "very important" or "quite important." In all countries but two, men are more likely to discuss politics "frequently" or "occasionally" (rather than "never") than are women.[4]

This is not to say that women have never been involved in political power seeking (Golda Meir and Margaret Thatcher immediately come to mind). Women of course do have power within kin groups and exercise informal but powerful influence in families. In those societies where women have formal political power, they often use it to increase the reproductive success of sons (Low 2000). Overall, on average, men find political activity much more worthwhile than do women and engage in it much more intensely. If we view the propensity for seeking political dominance as distributed on a continuum as a normal distribution, there is a distribution for males and for females, but the male distribution has a higher mean than the female distribution. In the upper tail of the distribution—the part including those most inclined to seek power—we will find many more males than females. Thus, most of the time, the most aggressive individuals in seeking political power will be men. In a large society such as most modern countries, only the most aggressive will succeed in rising to the top, and these will mostly be men because there will be many more males in the tail of the distribution.

Political power within an existing group is something of a zero-sum game: if one man has more, then he has power over someone else, who must therefore have less. The number of females in a group is also fixed, so if one man has more, then someone else must again have fewer (unless some can be taken from the group next door, a common strategy during evolutionary times, as discussed in chapter 2). In this sense, access to females and political power differ from wealth: as a society becomes wealthier, everyone can experience an increase in wealth. As a result, while men have always sought power, less dominant men have also sought to limit the power of those who are more dominant. This struggle has been going on at least since our ancestors have been humans, and probably long before. De Waal (1982) discusses similar struggles among chimpanzees. Much of written history is a history of this struggle at various times and places. For just one example, the American Revolution was (from a biological perspective) about one group of males, those in the colonies, seeking more power for themselves at the expense of another group of males, those in power in England. Telling the story as if this were the entire explanation is to trivialize many historical events, for, as we will see, the method of seeking power and the terms of power can also have important consequences. But this is at least part of many historical episodes, and to ignore it is also to tell an incomplete story.

In chapter 1, I discussed the nature of an evolutionarily stable strategy (ESS) and showed that at such an equilibrium differences between individuals might be present. Some individuals will play one strategy and some will play another. The number choosing a particular strategy will sometimes depend on the number of other individuals choosing each strategy; that is, the outcome is frequency dependent. This discussion is relevant for our analysis of politics. Individuals will generally choose that endeavor that maximizes their own fitness, given the choices of others and other constraints they face. Thus, some will choose politics as a forum, and others will make other choices such as business or academia. Those choosing politics will do so because they can do better in the political arena than in other activities; in economists' terms, they have a comparative advantage in politics. But politics differs from other activities in that it relies on power or coercion to get things done more than do other activities. Therefore, we might expect those individuals who are active in politics to be more ready and willing to use power than are others. This has important implications that I address below.

Political Power among Hunter-Gatherers

The story of human freedom begins in hunter-gatherer societies. In these societies of the sort that were predominant from the beginning of the existence of modern humans and probably even before, until about ten thousand years ago, a good deal of equality existed. While some men sought dominance, others in the group actively resisted this effort. This pattern has been studied most carefully by Knauft (1991) and Boehm (summarized in Boehm 1999). Boehm characterizes mobile hunter-gatherer societies as having a reverse dominance hierarchy, in which subordinate males form coalitions to limit the power of would be dominants. Resistance can take the form of simply leaving the area, because the people involved are often highly mobile. Other strategies include ridicule, refusal to obey commands, forcible resistance, and even homicide carried out against those with too strong a desire for power, those whom Boehm (1999) called upstarts. But the basic point is that among mobile hunter-gatherers, the social structure prevalent through most of our existence as humans and even before, society (at least among males) was highly egalitarian with respect to political power, although individuals have generally sought to become dominant. In this sense we may say that the desire for freedom and autonomy is a natural human desire and a part of the usual pattern (in terms of percentage of time of the existence of the species).

Mobile hunter-gatherers have relatively little social structure. For example, they do not seem to have much in the way of hierarchical structure, except for the reverse dominance hierarchy discussed above. Group sizes are quite small—typically about fifty to one hundred, with loose organizations of up to perhaps three hundred. Production technologies are quite simple as well, and there is little division of labor, except again by gender —males hunt, females gather—and perhaps by age. Carneiro (2000) indicates that each band and village was a sovereign unit, and each defended its sovereignty. Little political power is found in these societies. To repeat, this was the ancestral pattern for most of the existence of humans and most of the EEA.

Sedentary Societies

The next stage anthropologists identified is what is now called sedentary hunter-gatherers (see, e.g., Kelly 1995, for a discussion of the distinction.) The movement from mobile to sedentary hunter-gatherers marks a crucial distinction in human behavior. Increasing population pressure probably induced production methods that were less land intensive than foraging (e.g., Tudge 1998; Carneiro 2000). Sedentary hunter-gatherers and horticultural or agricultural societies are much more complex than mobile hunter-gatherers. Sedentary hunter-gatherer societies are much more like agricultural societies than like mobile hunter-gatherer societies. An intermediate stage is horticultural—with some farming, but with relatively simple technologies (Maryanski and Turner 1992; Kelly 1995; B. Smith 1995; Tudge 1998). Band and group sizes are much larger for agricultural societies, ultimately increasing to the thousands, or even more. There is substantial division of labor and greatly increased wealth and productivity.

Because of the sedentary nature of these societies, it is possible and worthwhile for them to create fixed structures and other forms of capital or wealth. In these societies, there is much more political inequality than among mobile hunter-gatherers. Often organized conflict approximating warfare occurs between groups, and armies are organized hierarchically, unlike the disorganized raiding common to mobile hunter-gatherers (Manson and Wrangham 1991). A breakthrough came with the invention of dividing males into age grades, with younger males forming something like a warrior caste. At this point war became more frequent, and the purpose became capturing land and engrossing villages for the purpose of tribute. This lead to increasing the size of chiefdoms and ultimately to the formation of states (Carneiro 2000).

A key difference between mobile hunter-gatherers and agricultural societies is population density. Small hunter-gatherer bands use relatively large amounts of land for subsistence. As population increases, settling down and engaging in cultivation, which produces more food per unit of land than does hunting and gathering, becomes desirable. Increased output in turn enables population size to expand even further; we have not yet reached the limits of this process. As population density increases, some states or proto-states will find aggression against neighbors worthwhile. This will lead neighboring states to also increase in size in order to defend themselves.

This is the balance of power competition Alexander (1987) discusses. As states become militarily powerful, dominant individuals in the state can obtain power over the residents of their own state as well. Perhaps this internal power is a byproduct of the predatory power. Once an army is created to conquer or defend against the tribe nearby, it is relatively easy for the army to dominate its own residents as well. An army is one source of the great political power of rulers and elites in agricultural states. An additional source is that successful predatory states often treat those who are conquered as subjects and may use their power to oppress them. Any history of any empire will tell this story.

Thus, in such societies, leaders often have substantial power and often use this power to engross for themselves many females (Betzig 1986), leaving many males with none. This behavior indicates that the aggressive egalitarianism of mobile hunter-gatherers was probably a good biological strategy. Sedentary hunter-gatherer and agricultural societies are what we think of when we think of early human history, though much more time as humans was spent during prehistory as mobile hunter-gatherers. They are also the beginning of states, which can become extremely large. In this stage, political power becomes most concentrated and uneven. Because these societies have included most humans during most written history, we tend to think that this is the normal course of human affairs. The works of Knauft and Boehm, in pointing out that humans have been relatively egalitarian through most of their existence, if not their written history, is quite important.

Polygyny and Monogamy

One extremely important change in human institutions with important implications for the analysis of political power is the invention of socially imposed monogamy—limiting each man to one wife. An important effect of this is the leveling of reproductive opportunity. That is, it is no longer possible for some males to engross excessive numbers of women (Alexander 1987). Alexander stresses that such reproductive leveling induces young males to make the effort needed to engage in warfare and is an essential part of the large modern nation-state.

A decision as to whether to allow polygyny is not an individual decision. It affects the entire group because the number of females is fixed. This is not an argument about group selection; I am not arguing that individuals will make decisions against their own interests for the benefit of the group. Rather, I am arguing that different groups may choose different rules and that some of these rules will turn out to be better than will others. That is, a group may choose some rule for any reason and there are many possible equilibria with respect to choice of social institutions or rules. However, if one rule is in some sense better, the group with the rule will increase in size at the expense of other groups, so the number of individuals subject to the more desirable rule will increase (Boyd and Richerson 1990). Then some groups will choose monogamy and some will choose polygyny; both are clearly viable choices, and societies have in the past and at present chosen both. More societies have chosen polygyny than monogamy. Polygyny is in the interest of powerful males, which is why it is so often chosen. Monogamy is in the interest of the less powerful males, who may not have wives in polygynous societies. The existence of monogamy or polygyny is also related to ecological factors, including parasite risk. Increased parasite risk provides a greater value for variability of offspring and so increases chances of polygyny (Low 2000).

From a biological or fitness perspective, groups benefit from polygyny— particularly in warlike societies. Victors can obtain additional wives from the conquered, thus benefiting their genes and leading to larger populations in the next generation. Moreover, some members of the victorious side will have been killed, and allowing polygyny means that the females (widows or those who would have married the deceased) will obtain mates. Losers will often also have a surplus of women (unless the victors take all of them) because of the loss of males, and allowing polygyny will enable these women to reproduce, thus increasing the size of the group. Indeed, Weisfeld (1990, 27) indicates that Muhammad instituted polygyny so that women could marry after the battle of Uhun left many men slain. Additionally, powerful males benefit from polygyny, even if it is not in the interest of the group. Thus, it is not surprising that most societies have been polygynous.

Groups also benefit from monogamy, however, as mentioned above. In particular, monogamy increases loyalty to the group, since all males will expect to mate. This may be particularly important in motivating young males, who are always the warriors in a society. By knowing that they will have mates, they may be motivated to be better fighters and more loyal. Knauft (2000) points out a sharp break in human societies between married and unmarried men or between males with and without legitimate access to females. Polygyny then increases the number of the latter, and, therefore, the social strains in a society. Moreover, the existence of polygyny means that some males can expect their lack of access to be a permanent rather than a temporary state, and this leads to additional conflict in a society. The substantial inequality caused by polygyny probably means that only monogamous societies could become modern democratic states. Of course, monogamy is not sufficient to generate democracy; the west became monogamous long before it became democratic. But it may be necessary.

For example, the level of inequality in wealth in the United States, as

Table 5.1. Share of Aggregate Income Received by Each Fifth of Families, 1994.

Lowest fifth	Second fifth	Third fifth	Fourth fifth	Highest fifth
4.2%	10%	15.7%	23.3%	46.0%

SOURCE: U.S. Census Bureau online, http://www.census.gov/hhes/income/incineq/p60tb1.html

shown in table 5.1, if translated into inequality in access to females, would be inconsistent with democracy. That is, as a crude approximation, if we take families as being males, the highest 20 percent of the males would have almost half of the females, or 2.5 wives per person. The lowest 20 percent would have about 5 percent of the females, so that 75 percent of these individuals would have no wives. This would create substantial social pressures, and probably would require a fairly repressive government to be sustained.[5]

Even today, Islamic states that allow polygyny have governance difficulties. The popular press has noted the relative lack of democracy in such states (see Crossette 1998). La Porta et al (1999) and Barro (1999) in academic studies also find that states with Islamic law are less democratic than others, after statistically controlling for other factors. Weisfeld (1990, 32) notes that in Arab societies there is a problem with "floater males," characterized as "desperadoes." He also indicates that disunity and feuding in Arab societies may be due in part to intense male competition, fueled by polygyny. It may be that nondemocratic, repressive governments are needed to maintain order in societies with large numbers of such males. This is particularly true since young males are likely to be without women in such societies, and young males are always the most volatile part of human societies and are associated with increased conflict (Mesquida and Weiner 1996). Mesquida and Weiner also point out that polygyny is associated with increased warfare. It is interesting to note that Turkey is one of the most democratic Islamic states, and polygyny is illegal in Turkey.

Polygyny is also relevant for analyzing terrorist activity. In a polygynous society, poor males will generally not marry or reproduce. If a male has no prospect for marriage, then from a biological or fitness viewpoint he has no future. If a person in such a situation commits suicide, then again from a biological or genetic perspective, there is no effect, since he would not have reproduced anyway. But if he commits suicide in a way so as to die a hero, his relatives (and hence his genetic contribution and inclusive fitness) may benefit. That is, having one martyr in a poor family may well create sufficient benefits for surviving members of the family so that their genetic fitness and the inclusive fitness of the suicide himself can increase. Lelyveld

(2001) indicates that families of Palestinian suicide bombers benefit at least by obtaining new apartments and furniture and that the bombers themselves are viewed as heroes. It is thus probable that we humans have evolved tendencies to be particularly altruistic to kin in situations where we as individuals cannot breed anyway. While we do not think of terror bombing as a form of altruism, it may serve this function with respect to the genetic relatives of the bomber who can benefit from this behavior.

Polygyny would seem to benefit women in that it provides more alternatives. A woman can become the second or third wife of a successful male, rather than the first wife of a failure. Through normal principles of supply and demand, this increase in demand should translate into increased value for women. For example, men would be expected to compete for women by offering better terms of marriage, such as increased freedom. But in fact, we observe the opposite in at least some polygynous societies. For example, in Islamic societies women seem to have fewer rights and are often kept in seclusion. The answer may be that the number of unmated young males creates some danger of illicit sexual activity by females (either voluntary or through rape), and their husbands respond by excessive mate guarding—keeping women isolated and away from such males. It may also be that male relatives appropriate the increased value of females, and polygynous societies often have substantial payments (bride price) for females paid to their male kin (Weisfeld 1990). Additionally, in polygynous societies families invest more in males because of the potentially higher payoff (Low 2000). Hrdy (1999) points out that a man with several wives will gain in fitness terms even if some of his children die, as long as he can sufficiently increase the number of surviving children by adding additional wives. Since women are limited in the number of children they can bear, this will not benefit them.

Throughout much of the history of the west, not only was marriage monogamous, but it was also permanent. More recently, divorce has become common and some men marry several times. Some wives are significantly younger than their husbands; indeed, the term *trophy wife* has been introduced to describe this practice. This raises the following question: If some men marry more than one young wife during the period when the wife is fertile, then some men will not marry fertile women. That is, if some men marry more than one wife during the period of peak fertility, some men will not be able to father any children (see Wright 1995). Siow (1998) discusses some implications of differential male and female fecundity. One implication of his analysis is that in a system where divorce and remarriage are allowed, poor men (in his terms, men who are unlucky in the labor market) will never marry. Such men would be better off in a system with no divorce. Why has this increase in allowing of divorce and remarriage not led to the inequality and social unrest or control associated with other forms of polygyny?

I believe that the answer is in three parts.[6] First, society no longer condemns homosexuality, and male homosexuality is more common than female (Posner 1992). Thus, some males are withdrawn from any contest for females.

Second, there are simply more females than males today. This has been true in the United States since the 1950 census. Before that, there were consistently more males than females.[7] This is not due to changes in gender-specific life expectancy. Even with the relatively high maternal death rates before the 1940s,[8] females have had longer life expectancy than males since at least 1900.[9] It may be that increasing overall life expectancy is responsible for this change. More males than females are born, but males begin dying in numbers in excess of females immediately. As a society ages, the surviving females will come to outnumber males.

Moreover, a slight surplus in the number of females can translate into a large surplus in the number of unmarried females if most heterosexual persons are married. Think of a population of one thousand persons, 510 females and 490 males. Assume that there are 480 married couples. Then there are forty unmarried persons—thirty females and ten males. A 2 percent difference overall in the number of females has translated into a three-to-one majority of single females relative to males.

Third, and related to the second point, modern technology has enabled females to remain sexually attractive for a longer time, so that males may be satisfied with relatively older females. Natural selection has caused males to desire fertile females as sexual partners. However, natural selection selected for certain features associated with fertility, rather than for fertility itself. Thus, features such as a low waist to hip ratio, clear skin, firm muscles, and hair that was not gray were associated with youth and fertility. Males who preferred such features would mate with fertile females and those males left more descendants than males with other tastes, so they became our ancestors.

Modern technology, in the form of estrogen replacement therapy, cosmetics, and plastic surgery, and also reductions in number of children ever born to each woman, has enabled older females to mimic youthful appearance, and so males are attracted to these females. Indeed, Abed (1998) even argues that young females feel obligated to compete with older but youthful appearing females and that this competition is the cause of eating disorders of young females, such as anorexia. The percentage of all marriages (including remarriages) involving women over thirty five has increased from 13.8 percent of all marriages in 1980 to 21 percent of all marriages in

1990.[10] In 1995, there were 63,000 facelifts and 208,000 other smaller surgical procedures aimed at increasing youthful appearance; most patients were female (Gilman 1999.) Thus, the increasing number of older females in society are perhaps sufficiently sexually attractive so that the fact that some males marry several times does not lead to unhappiness and discontent on the part of other males. While these marriages may not lead to children, we are not selected to desire children, but rather to desire sex and then to love and provide for the resulting children.

Substantial gender imbalances in much of contemporary Asia are caused not by polygamy but by preferences for males and the possibility of gender-based abortion or infanticide (e.g., Dugger 2001). For example, in China, 117 boys are born for every 100 girls; in India, there are 1,343 boys per 1,000 girls from birth to age six. If the theory proposed here is correct, then these imbalanced sex ratios and the associated surplus of males do not bode well for democracy in these countries as the population ages and the excess males enter adulthood. It should also be noted that a decision to have a male child in a society with more males than females is a decision counter to fitness maximization, since the male is less likely to reproduce than would be a female. This is a case where cultural preferences outweigh fitness concerns.

Political Power in Modern Western Societies

Modern western societies are based on industrial production, rather than on agriculture. (We might be in a postmodern age based on information technology, but the characteristics of this age are not sufficiently known to analyze here.) This occurs because of increasing technology. Technology in turn increases because of two factors. First, increasing population size means that there are more individuals to invent new technologies (see J. Simon 1981/1996; Kremer 1993; and Jones 2001). Second, the invention and protection of property rights in intellectual capital induces more investment in creation of such capital (Jones 2001).

Power in modern western societies is much more equally distributed than in the agrarian societies that replaced hunter-gatherer bands. Conventionally, this is thought to be true because modern governments have restricted powers, and this is important, as discussed below. But there are additional significant reasons for the increasing equality of power because several aspects of modern life and technology serve to increase individual power and reduce the possibility of coercion by powerful dominant individuals.

Modern humans have basically two systems of political structure (Richerson and Boyd 1999). First is the evolutionarily old system ultimately based on kin selection and reciprocity between individuals who know or are related to each other, as discussed in chapter 1. This can support close ties and extensive cooperation, but only on a small scale, of perhaps ten to twenty-five individuals. Different mechanisms support large organizations and are less based on our most basic evolved tendencies. These mechanisms are unique to humans. These larger mechanisms do make some use of our evolved tendencies, as when states try to mimic the family, as discussed by Johnson (1986), which leads to what Richerson and Boyd (1999) call symbolically marked group membership—for example, national flags. We have evolved mechanisms that allow support of such larger groups. Indeed, Richerson and Boyd (1999) hypothesize that the crucial step in the evolution of modern humans was the ability to associate in larger groups and that Neanderthals may have lacked this ability. Moreover, the larger social agglomerations make use of the cooperative tendencies that evolved for cohesion in smaller groups. In what follows, I first discuss individual power with respect to the small groups similar to the groups in the EEA. I then discuss overall social power.

Small Groups

Recall first that societies in the EEA were small. Bands might have been as small as twenty-five individuals, with perhaps ten to twenty adult males. In these bands, there was only one hierarchy (which Boehm and Knauft indicate was a relatively weak hierarchy). That is, a dominant male would generally dominate a dozen or so subordinates, with the extent of domination limited by the factors discussed by Boehm. This is the evolutionary pattern, and it still echoes in humans today. For example, we may be selected to compete with a relatively small number of other individuals for dominance. We seem to create work environments similar to those in the EEA even today because such environments are consistent with evolved tendencies and, therefore, work more efficiently.

For example, in my day-to-day life, I interact most closely with the fifteen to twenty members of the economics department at my university. Even though this department is embedded in numerous hierarchies (the social sciences division, the college of arts and sciences, the university), for most purposes these hierarchies have little influence my life, and I can go for weeks or months without thinking much about them.[11] I have also worked in three separate parts of the Federal government,[12] surely one of the largest hierarchies in the world. Nonetheless, even in this environment, I was part of relatively small work groups. In managerial positions, I was in charge of my own staff (e.g., the fifteen to twenty economists at the CPSC) but was a lower ranked member of the senior staff of that agency, subordinate to the chairman and some others. That group of senior managers was also about fifteen people, so again, the evolved group size was invoked. I did not feel like a subordinate to the president, or to any other high officials in the government, though I of course was, because the gap was so wide that I never had any interactions with these high officials and so felt no personal subordination.

I do not think that this pattern is atypical: I think that most people are in working or social environments that are not dissimilar in size to the small group size environments of the EEA. The military organizes itself in a similar way, and those military organizations that best simulate environmental conditions similar to those in the EEA seem to be the most effective (Richerson and Boyd 1999). More generally, organizations and societies that organize themselves in this way are more successful than others. As Richerson and Boyd (1999) point out, one method of improving the functioning of organizations is selection, but humans can also plan and design mechanisms to achieve their goals.

Because humans are evolved to desire autonomy and freedom from coercion in relatively small organizations, such organizations have advantages over others. First, people may be more productive in this form of organization, since the structure makes use of evolved tendencies to interact with relatively small groups. Second, because people prefer such working environments, they will accept relatively lower pay to work in these environments; alternatively, they will demand higher wages to work in different—for example, more coercive—environments. Because western capitalist societies are highly competitive, there are strong pressures to design efficient organizations. Thus, there are pressures to design relatively free non-coercive work environments with relatively small work groups.

As an alternative, consider the former Soviet Union. This was a much more structured society, and coercion was much more common at all levels. The result has been that in a sort of Darwinian competition, this structure lost out. While coercive societies may have been feasible during some stages of technology, modern technological systems do not do well with such coercive systems. Capitalism is a system of freedom and also a highly productive system. Productivity enabled the west to defeat the Communist countries, but this productivity is only possible in a free society. This is an important explanation for the increasing freedom that exists in modern societies as compared with the coercive empires of the last ten thousand years in agrarian societies. Boehm (1999) and Maryanski and Turner (1992) also discuss increased freedom in modern democracies. One feature of the modern world leading to increased autonomy is that there are pressures for individuals to operate in relatively small work groups, which operate with as little coercion as is feasible.

One difference between dominance hierarchies in the EEA and today is highly important. In the EEA, there was a single dominance hierarchy with respect to all aspects of life. The group would probably be an extended family, a work (perhaps hunting) group and a military group with respect to raids on neighbors. The same hierarchy would prevail in all spheres because the same individuals participated in all spheres and engaged in all group activities. Thus, an individual who was a subordinate was a subordinate in everything, and similarly for a dominant.

This is no longer true. Today, humans can belong to many different groups containing many different individuals. Humans join organizations in addition to work groups, including religious institutions, clubs, and other voluntary groups. Again, these tend to organize themselves into relatively small subgroups. Individuals who are subordinate in one setting can be dominant in another. A person who might be a subordinate at work can be president of a club or in charge of a subcommittee within a church group.

People also are members of families, and family structure is quite important to most individuals for evolved reasons discussed earlier. Indeed, more people view family as very important than any other factor.[13] In most western societies, we no longer live in large extended families; rather, the nuclear family (husband, wife, children) is the most important family structure. One effect of this pattern is that adults are not subject to control by older family members, as is true in extended families. This freedom may be particularly important for women because in a patrilocal society women would typically live with in-laws and would be relatively powerless in such a setting (Hrdy 1999.) In some societies this pattern still persists (Weisfeld 1990).

The result is that people generally belong to a noncoercive hierarchy at work and one or more additional groups, including a family. For example, in the United States, between 70 and 80 percent of individuals belong to some group, and 5 percent belong to six or more organizations (Baumgartner and Leech 1998). People who seek some amount of dominance can often join a group in which they can obtain some authority because the number of groups is large and potentially limitless. In many voluntary groups it is difficult to induce individuals to take leadership positions, and almost anyone who wants to be a leader can become one in some environment.

Moreover, the increasing productivity of capitalism means that most people in western societies have access to wide ranges of goods and services. We can consume things that were not available to our ancestors—even to the most powerful of them. While this ability to consume is not generally thought of as a form of freedom, it provides people many consumption options not available in the past; in this sense, it is indeed an important dimension of freedom. For example, in feudal times travel was forbidden to many and difficult for all, and freedom to travel is an important component of freedom. Now we are allowed to travel. But we also have the means to travel. A random person in the United States can travel faster and farther than could even a king in past times. This is but one example of the increasing freedom associated with the wealth capitalism creates.

Governmental Power Today

In addition to the small groups in which we work and live, there is also overall power in the society. Conventionally, this power is thought of as being associated with government. Government is quite powerful in the United States and other western economies, but several factors actually serve to limit the power of government.

First is the system itself. The U.S. Constitution was consciously designed to limit the power of government; this is the system of checks and balances that we all studied in school. The Constitution is consistent with evolved preferences (Grady and McGuire 1999). This system does really create several limits to government power. For example, individuals have the right to challenge in court many actions of the government. As I write this, Microsoft is in the process of appealing its conviction by the government for antitrust violations to a higher court and the survivors and heirs of the Branch Davidian deaths in Texas are suing the United States government for its actions there. In the United States, such challenges are routine. Under the sorts of governments that prevailed during agrarian times or in communist and other dictatorial systems, such challenges to the power of the government itself would have been impossible.

Second, and perhaps more important, in a modern society government is really only one player among many and often not the key or most important player. Capitalism creates an unlimited number of hierarchies, so there is no need to compete only in the political hierarchy. For example, CEOs of many large companies make significantly more money than anyone in the government, including the president, and can probably spend their money with more freedom and less publicity. Many entertainers and sports figures are as well known as major government figures and again can make more money and have greater freedom in spending. Rumor has it that such figures also have access to numerous females (see Miller 2000 for a discussion). As recent events in the United States have shown, government authorities may have less such access. In agrarian societies, the most important route to wealth and power was government. That is not true in the west today; many other channels are available, and anyone who wants to enter a competitive hierarchy has numerous options.

Of course, government has ultimate power over businesses and individuals and can bring the force of the law against a business, no matter how successful the enterprise might be. The prosecutions of Microsoft and of Michael Milken (Fischel 1995) demonstrate clearly this power. But the number of such prosecutions possible at any given time are quite limited because of the legal protections and the associated money and time cost to the government of bringing the prosecution. Moreover, the government must have at least some legal excuse for bringing such prosecutions; the decision cannot be completely arbitrary. Most successful entrepreneurs need not fear any penalty from the government. Additionally, the government is not guaranteed to win if it does bring such actions. The system of checks and balances also means that even if government wins in one forum (say, the courts) losers can use another forum (say, the legislature) to overturn the victory (Rubin, Curran, and Curran 2001). Thus, the government has ultimate power, but even this power is clearly circumscribed.

Capitalism generates wealth to combat and limit government. Successful firms can spend money to lobby and also promote their positions, which serves as a partial check on a democratic government. For example, it was my experience at the Federal Trade Commission that many people would leave the agency to work for the private sector, often in the role of attorneys or consultants representing firms being challenged by the FTC itself.[14] Private work simply pays better because citizens are unwilling to allow government salaries sufficiently high to attract and retain the best people. As experienced people leave government for more money, the power of government is limited because these people often work to limit the power of government.

Indeed, one of the great human achievements of the recent two to three hundred years has been reducing the power of government relative to other institutions in society. Given the innate human preferences for either being a dominant individual or for avoiding being dominated, a powerful government can benefit a few individuals at the expense of many. This has been the pattern throughout most of the time of recorded history. Moving from this pattern to a system with severely limited government powers has, therefore, caused a major improvement in human happiness. For example, democracies are significantly less likely to engage in genocide and other forms of state sanctioned murder than are other forms of government (Rummel 1994). Moreover, the level of such murder also decreases with the wealth of the society (Scully 1997).

Other institutions such as large businesses have some power in societies, and this power may come at the expense of government or of individuals. But there are significant differences. First, businesses do not have access to coercive power in the form of violence; only government has this power. Second, individuals have some ability to determine if they will be subject to the power of business—even of monopolies. If I want to use a computer, I will probably use a Microsoft product—but I do not need to. I could buy an Apple, or even a typewriter. I might work for a large company, but the number of employers is huge and none has monopsony (demand side monopoly) power. But I cannot escape the power of the United States government unless I emigrate, and even then I will become subject to some other government. Third, even if some business has power for a time, competition and entry of other businesses is possible and indeed is a recurring event. The very profits a monopolist earns are an incentive for others to enter the business (unless government itself keeps them out, as is common in public utilities). But the only way a competing government can come into being is through a revolution—an uncommon and difficult task. Political parties add some measure of competition, but the underlying institution is stable and entry is difficult or impossible.

Government must have access to force. The role of government is to create public goods—goods that benefit everyone in society whether they pay for the goods or not. The major example, consistent with the arguments in this book, is defense. Humans are and have always been a predatory species, with the main targets of predation being other humans. Government then serves to protect its citizens from this predation or prevents them from engaging in such predation against neighboring groups. But in either case (and especially in the case of defense) individuals benefit whether or not they pay. The only way to induce them to pay is through a system of taxes, collected through the threat of physical coercion and violence. Another public good with a long evolutionary history is the provision of law and order and contract enforcement within a society, the legal system. Again, this requires violence or the threat of violence, both to finance the provision of the good and to enforce its decrees. Public health is also a public good, because contagious diseases create negative externalities, but it is one perhaps without such a long evolutionary history. Enforcing mandatory monogamy laws may also be a public good.

But once government is given the right to use force, government must restrain itself. This is a difficult project, and many human societies have failed in this task. Societies that cannot effectively restrain government are dictatorships. In the EEA, societies were sufficiently small that coalitions of nondominants were able to form reverse dominance hierarchies and restrain the poser of would-be dominants. Throughout most of recorded history, such restraints were lacking and humans were subject to powerful governments. More recently, with the advent of true mass democracies, we have again been able to create restraints. But we must remain vigilant; the danger of even democratic governments reverting to dictatorship and coercion, as the example of Nazi Germany shows, remains. The answer, as Nobel laureate in economics Friedrich Hayek (1944) indicated long ago, is to restrain the power of government and limit it to those functions that are truly public goods.

There is another reason for limiting power of government. I argued above that different individuals have different strategies for seeking to improve their own fitness. In Reiss's (2000) scheme for classifying individuals, one dimension is power. Arnhart (1998) identifies political rule. In a modern society with open choice of individual occupations and roles, some will choose business and some government; some of us even choose academia. But presumably each person will choose that occupation where he or she expects the maximum returns in terms of utility or fitness. Therefore, those choosing government on average will expect to do better by using coercive power than by using other forms of influence. In other words, we would expect politicians and perhaps senior government bureaucrats to be those members of society who benefit most from using power to achieve their goals.

The fact that elected officials are selected from among those who seek power explains another puzzle in modern politics. In contemporary America, political liberals, generally associated with the Democratic Party, seek to regulate economic behavior and economic competition. They generally favor higher taxes (which restrict the ability of individuals to spend their money) and more controls over business. Of course, controls over business also translate into controls over individuals. If the Food and Drug Administration (FDA) does not allow a pharmaceutical company to sell a product, then no consumer can buy it. If firms are not allowed to pay less than a minimum wage, then workers are not allowed to work for less than this amount. Some basis for this desire for regulation may in our tendencies for envy, as discussed in chapter 4.

Conservatives, generally associated with the Republican Party, seek to control consumption activities of individuals. They are generally in favor of restrictions on behavior, such as regulation of sexual behavior, abortion, and pornography. (Traditionally, they might have regulated gambling, but since government has entered this business through state lotteries, and since various legal rulings have allowed native Americans to open many casinos, this issue is less important.) In chapter 6, I discuss the basis for this desire to regulate personal behavior.

But the puzzle is that neither party is in favor of overall freedom. That is, there is little constituency in the United States for a political position that would regulate neither economic nor consumption behavior. (The exception is the Libertarian Party, but this party typically does not do well in elections.) The explanation may be in the power-seeking behavior of politicians. If those attracted to politics use politics as a method of seeking status and if they were selected to be those in society best at using power, then

there would be relatively little demand for positions associated with elimination of political power. That is, those who might seek to reduce the power
of government in all dimensions would tend not to seek political power in
the first place. Given this, those of us not involved in government would do
well to form our own reverse dominance hierarchy and attempt to limit the
power of government.

In the short run, some may find this costly, as they might want to use
the power of government to advance their own ends; some individuals or
groups (special interest groups) are sometimes able to harness the power of
government for their own interests. Economists have analyzed this as rent
seeking (e.g., Buchanan et. al. 1980). Those in a position to benefit from
government power may advocate the growth of such power. But the long-
term effect will be to increase government power, and this generally serves
to harm individuals. In the long run, all will benefit from reduced power of
government.

Of course, this incentive to use government power for a private benefit is
an example of a prisoner's dilemma, of the sort discussed at several points
in this book. The best solution for such a dilemma would be to agree to limit
the power of the government, so it would not pay for any group to even try
to harness this power for its own benefit. Some believe that the original U.S.
Constitution was written to achieve exactly this goal and only in relatively recent times (since the 1930s) has this agreement broken down and the power
of the government been made available for private parties. Buchanan and
Tullock (1965) is the classic statement of this position.

I find it puzzling that so many in society advocate stronger and more
powerful governments. Of course, government itself through politicians
and government employees would benefit from greater power, and it is not
surprising that government agents and employees would seek additional
power. But it is surprising that so many others, and so many intellectuals,
also seek to increase the power of government. (Maryanski and Turner
1992, also express puzzlement at intellectuals'—in their case, sociologists'
—dislike of the modern industrial capitalist environment.) It is particularly
surprising that feminists are often critical of western society for, as Hrdy
(1999) points out, women in modern western societies have much more
power and freedom than has ever been true before.

The answer may be in the forces identified in chapters 3 and 4. In chapter 3, I argued that humans have some altruistic tendencies. Some who
advocate stronger government may perceive government as a source of altruism and may want to strengthen its position for that reason. This is a viable position, but we must realize that there is a tradeoff. As government
becomes more powerful and engages in more redistribution, national income and wealth will fall. One of the most powerful methods of increasing

the well being of the poor is a strong economy, and government interference can weaken this economy if it goes too far.

A second reason is that identified in chapter 4, where it was argued that many today confuse the roles of productive and dominance hierarchies. To the extent that there is confusion, many persons may view business firms as primarily dominance hierarchies and not perceive that they are in fact productive hierarchies. If this confusion exists, then individuals might want to use government to restrain business hierarchies. The fallacy in this argument is that ultimately government is more of a dominance hierarchy than business can ever be, because business depends on obtaining the voluntary consent of its clients (customers or workers), while government can obtain this consent through force.

A third explanation, discussed more fully in chapter 7, is that we are not good at detecting the effects of behavior; we are better at observing motives than results. As I discuss more fully later, this means that many people are hostile to business and to capitalism, because capitalism is based on selfish motives. The point that many miss is that these selfish motives often lead to socially beneficial outcomes.

Singer (2000) illustrates this point. He argues for a Darwinian left, and I find some of the evolutionary structure of his argument compatible with my beliefs, even though my own predilections are not for what is normally called the left and I do not agree with many of the policy implications he derives from his analysis. But Singer also commits serious errors. He suggests that one way to improve society is to substitute cooperation for competition, and he views the market as the essence of competition. In fact, the market is the most efficient mechanism humans have developed for cooperation. A large firm employs thousands of individuals who cooperate with each other in producing the output of the firm. But this is only the beginning. The firm buys inputs through the market from other firms—another form of cooperation. The firm sells its products to others and ultimately to consumers. This is another form of cooperation. Singer (2000, 51) urges that we use the results of game theory to "make it possible for mutually beneficial relationships to flourish where otherwise they would not." But that is exactly what the market does: it fosters mutually beneficial relationships between owners of capital and labor and between buyers and sellers.

Indeed, the internationalization of the economy means that all of us cooperate with most of the other humans on earth, but the cooperation is mediated through the impersonal market and we do not observe it directly. I may buy a shirt made in China. The worker who sewed this shirt may use a machine made in Italy, and he depends on Chinese agriculture for food. In this way, I have an economic link with the Chinese and the Italian economies and the producers in them. Similarly, I sell my services as a professor to students, many of whom (particularly graduate students) are from countries such as China and India. My particular university, Emory, derives much of its income from Coca-Cola, and this company sells much of its product in other countries. The market provides a method of cooperation to literally billions of individuals, even though the cooperation is indirect and most of us do not see or meet each other.

The analysis of political power tells us that humans are selected to seek it whenever possible, and this would apply to government agents as well as to others. But it also tells us that those who could expect to be subordinate should resist the efforts of dominants to increase their power. We should apply this lesson more strongly to government.

Summary

Freedom has been an important characteristic of human political systems throughout most of our existence, and humans have a strong taste for freedom. On the other hand, human males and prehuman ancestors have sought political power as a way of obtaining access to females. But humans have also tried to avoid being dominated. The best evidence is that through most of human existence, when our ancestors were hunter-gatherers living in small bands, dominants or would-be dominants had relatively little power. This was not because they did not seek such power, but rather because nondominants banded together to limit the power of the dominants. With the beginning of sedentary societies and of agriculture, the power of dominants increased substantially. Most of written history is the story of conflicts between various groups of dominants or between dominants and subordinates. Until relatively recently, dominants generally won.

A necessary condition for the increase in democracy and the reduction in the power of dominants was probably socially imposed monogamy. Polygynous societies have some advantages in competition with other societies. However, they also have costs. In contemporary conditions, the advantages of monogamy seem to outweigh those of polygyny. In any event, polygynous societies create bands of unmarried young males, and control of these bands may require a coercive state. Evidence indicates that Moslem societies, which allow polygyny, are less democratic than others.

Modern western society limits the power of dominants, and individuals have more freedom now than at any time since our ancestors were hunter-gatherers. Indeed, because women are free and because of the additional consumption opportunities created by capitalism, we in the democratic west have more freedom than humans have ever had in the past. Several

forces, including competition from other hierarchies, constrain government. There are substantial benefits form limiting government power and
great dangers from allowing it to increase. It is, therefore, a puzzle to explain why so many seek to increase the power of government relative to
other institutions in our society, although some of the arguments in this
book can shed some light on that puzzle.

CHAPTER 6

Religion and the Regulation of Behavior

This book is about politics. But religion is an important part of political behavior. Religion interacts with politics in several ways. As discussed in chapter 2, religion is one way in which humans define their identity when engaging in group behaviors, including conflict. Many religions also foster cooperation, at least between coreligionists. On the other hand, religious wars have been a prominent part of human behavior for as long as we have records. The Old Testament describes such wars in detail. The Crusades are another example of a religious war. Moreover, as I write this, conflicts, from time to time, erupt as wars between India and Pakistan; between Serbs, Albanians, and Croats in the former Yugoslavia; between Protestants and Catholics in Ireland; and between Jews and Moslems in Israel. The September 11, 2001, attack on the World Trade Center was viewed by the attackers as part of a religious war. As discussed in chapter 2, none of these conflicts are ethnic conflicts; all may be better characterized as religious.

In addition to its role in defining group identity for conflict, religion plays roles in political behavior. In the United States today, one of the most contentious political issues is the regulation of abortion. This question has important additional implications, such as the use of stem cells in medical research. Abortion is an issue that makes no sense as a political issue except that it is fueled by religious belief. Other topics are based on religion or religious beliefs. Some examples include: some regulation of speech, including the attempts to regulate pornography on the Internet; issues dealing with homosexuals, such as the legitimacy of gay marriage; regulation of gambling, prostitution, and drugs. Thus, religion is an important part of political behavior, and a book on the biology of politics would be incomplete without a discussion of religion.

Sources of Religion

I do not take a position on the validity of religious belief. However, from a scientific position, explaining religion is clearly puzzling: religious beliefs are nontestable. That is, one must accept them or reject them on faith; there is no objective way of determining if they are correct. Nonetheless, people spend large sums of money and effort on religion. That is, humans are willing to spend substantial resources on nontestable beliefs. It appears that people could have objectively better lives if they spent the resources on things for which a return is observable. This is the puzzle.

The major source on the evolutionary basis for religion is Boyer (2001) (see also Guthrie 1993). They make the following argument: The human brain evolved mainly to deal with other humans. (This is the position I advocated in chapter 1.) When something (anything, good or bad) happens, it might be due to purposeful action on the part of an intelligent agent or it might be the result of random forces. In deciding whether intelligent action is the cause, two possible errors can be made (the classical errors in any decision problem, studied in statistical decision making). We could decide that the action is due to purposeful intervention when it is actually random, or we could decide that it is random when it is really due to intelligent intervention. The second type of error is much more costly than the first, particularly if we can quickly correct mistakes. That is, if something is done purposefully and we assume that it is random, we might lose a substantial amount. If something is random and we decide it is purposeful, the losses will be relatively smaller. Therefore, we would, on average, do better if we assume that actions result from purposeful behavior because the losses of being wrong in this case are smaller. We have evolved to think in just this way.

This is a variant of Pascal's wager, the argument that the cost of not believing in God if there is a God (this cost is eternal damnation, which is an infinite cost) is much greater than the cost of believing in God if there is no God (which is finite). Therefore, this is an argument that one should believe in God. However, Pascal's argument is prescriptive, in that he is advocating a position. The argument Boyer and Guthrie made is explanatory—they claim that the human mind has evolved in the manner they describe.

For example: When meat is shared from the hunt, I don't get my fair share. It may be the luck of the draw. But it may also be that my neighbor has been spreading untrue rumors about me, and so I am being subtly punished. If it is just random, then next time I should get my share. But if it is my neighbor, then next time I will also be shortchanged. If it is luck, there is nothing to do. If it is my neighbor, I can try to placate him. The risk of wasting some time being nice to my neighbor when the cause of my shortfall was bad luck is relatively small; the risk of doing nothing when it was my neighbor is that I may starve. Therefore, it is rational to assume that my neighbor has been harming me and to placate him. We have evolved to believe in this way—that is, our predecessors who assumed luck when it was actually a neighbor's actions were less likely to survive and be our ancestors than those who assumed it was a neighbor when it was actually luck.

To extend the story to religion: A hunt is unsuccessful. It may be simple bad luck: I went east, the animals went west. It may also be because I did not propitiate the Hunting Gods by giving them the liver of the last animal I killed. If it is just bad luck, and I give the Gods the next liver, I am out one liver for no return. But if it is the wrath of the Gods, and I don't give them the liver, I will continue to fail in my efforts. It is then rational to give the liver to the Gods, even if it in fact has no effect on my success.

We have evolved, therefore, to believe that most actions effecting us are purposeful. That is, we have evolved to be highly anthropomorphic—to see intelligent human-like action and purpose throughout nature. More specifically, those of our predecessors who assumed that most events had purposeful causes were more likely to survive than those who assumed that most events were causeless and random. In Guthrie's analysis, religion is the result of this anthropomorphism. That is, we assign intelligent causes to what are actually natural events; we call these intelligent causes gods. Thus, according to this theory, people are religious because they expect and seek purpose in explaining events. More risk averse persons are more religious, which is consistent with this theory (Miller and Hoffman 1995.)

Guthrie discusses numerous other ways in which humans engage in anthropomorphism. For one example, we treat vehicles as being animated and refer to them with personal pronouns or name them after animals (Mustangs, Cobras, and others). For another example, Guthrie's book has a gallery of illustrations from advertisements of inanimate objects, such as coffeepots, depicted with human faces and features. We name hurricanes after people. An example from my field (not mentioned by Guthrie) is Adam Smith's invisible hand metaphor ("Every individual . . . is . . . led by an invisible hand to promote an end that was no part of his intention" [1789/1994, 485]). Economists have extended this metaphor to refer generally to the market, and we attribute wisdom and other desirable characteristics to the blind working out of forces in this market. Biologists sometimes speak of evolution and adaptation in similar terms.

This tendency to anthropomorphism explains why many people believe in some religion, but it does not explain the particular form of religion. Religious beliefs have varied tremendously over human existence, from simple animism to sophisticated theologies. Since all of these religions have existed, all are consistent with the evolved basis for religion. Moreover, because none are testable in the scientific sense, we cannot use science to discriminate between them in the way that we can decide on the heliocentric Copernican astronomical system instead of the old earth-centric Ptolemaic system or to determine whether acquired characteristics are heritable. Boyer suggests that religious beliefs have evolved through an invisible hand mechanism, so the remaining set of beliefs is consistent and well adapted and appears planned. However, he does not distinguish between belief systems that have great internal power and those that help or harm their believers. As a result, some apparently highly counterproductive religious beliefs have persisted in some societies. Edgerton (1992) describes in detail many of these belief systems. If all belief systems were counterproductive in this way, perhaps the tendency for religion could not have evolved.

We do observe that the religions in which people believe change over time. Moreover, the number of major religions has decreased over time, as adherents to some religions have become quite numerous and other religions have died out. These successful religions are mainly Christianity and Islam. (Hindus are numerous, but this religion differs from the two former religions in that it does not now proselytize and has grown mainly through the natural increase in the population of its adherents.) The best theory to explain this success is a model of cultural evolution, in which some cultural institutions have advantages over others. Boyd and Richerson (1990) have proposed a nice model of this sort. In this model, there are many possible equilibria, and the equilibrium reached in a particular population depends on the initial state of that population. Under many circumstances, each individual in a population will do best by playing the most common strategy in that population, but different populations will have different strategies because of different initial conditions. Competition among groups, however, can lead to one of these equilibria becoming more common than will others if it is more successful in meeting the goals of its adherents. This is not a group selection model in the sense that the term is generally used, for it cannot be assumed that individuals are forgoing private benefits for the good of the group. Rather, each individual is privately maximizing his or her own interests, given the state of his own group, but the result is that a strategy that is in some sense best will prevail.

Christianity and Islam, the surviving major religions, have some features in common. First, as discussed in chapter 2, these religions are not ethnically based, as were many earlier religions. Some religions—Judaism, Hinduism—do seem to be based on ethnicity, but the major religions, Christianity and Islam, are not (Hinde 1999). They accept and actively seek converts from all ethnic groups. They have obtained converts forcibly as well as through persuasion. This has been an important factor in the growth of these religions (Stark 1997). Moreover, as discussed in chapter 2, creation of group identity with religion rather than ethnicity as a basis has probably led to reduced levels of violence, though many conflicts over beliefs within and between the major religions exist. Today most religious persons interested in converting others seem willing to rely on persuasion, but some Muslims still believe in force and violence for conversion (Pipes 2001).

The other feature of surviving religions is that they dictate certain moral

behaviors, and these behaviors have often made their adherents more successful. This is because part of the benefits of these moral behaviors is to
solve prisoner's dilemmas and, therefore, lead to more cooperation among
members than would otherwise occur. They may be called efficient ethics,
in the sense that efficient altruism was defined in chapter 3. Note that individuals do not follow moral codes to provide these group benefits. Rather,
believers follow the dictates of their religion because they believe it is their
long-run interest, where the long run may be defined as including the afterlife. But those religions that are more likely to impose certain moral standards on their members are more likely to have successful adherents, and
those are the currently surviving religions. Evidence shows that religiosity
correlates with many social behaviors, including criminal activity; consumption of alcohol and drugs; health; and fertility (Iannaccone 1998). More religious behavior is associated with the good side of these factors and less
religious behavior with the bad side.

Even some early religious beliefs served to create efficient institutions.
Burkert (1996) discusses the role of religion in enforcing property rights
through treating boundary markers as being sacred. He also discusses the
role of oaths in enforcing contracts and promises. An oath would generally
call for retribution from the gods for violation and, therefore, could serve
as an enforcement device for believers. It is generally difficult for humans
to pretend beliefs that are not real (Frank 1988), so if appearing to have a
religious belief is valuable, it would pay to actually have such a belief. If
people were more likely to transact with someone who actually believes
in divine punishment for breaking an oath, having this belief would be individually valuable, because it would increase transaction possibilities. It
would be costly for a nonbeliever to pretend belief, so religious behavior
could act as a signal.

One could demonstrate a religious belief by contributing resources to
the gods or religion, for example, by financing sacrifices or, today, contributions to church, mosque, or synagogue building funds. Even if expenditures on religion are wasteful because the gods really don't care, they might
nonetheless be efficient for an individual if they increase value of transactions sufficiently to pay for themselves. For an individual, such spending
would be a signal in the sense of Spence (1973). A true believer would gain
both a valuable reputation as a believer whose oath could be accepted and
(he would think) whatever religious benefits are promised for such contributions. A nonbeliever would get only the reputational benefits and
would not expect any afterlife benefits. Therefore, only believers would
contribute. But because belief would solve the prisoner's dilemma, there
would still be net benefits from belief, even if the belief were not objectively
justified.

The argument is that religious beliefs originated for essentially the anthropomorphic reasons Boyer (2001) and Guthrie (1993) discuss. The form of these beliefs was arbitrary. At some point, some religions began advocating policies that led to increased productivity of their members. Then members of these religions began to do better than others and so these religions grew. This growth could have been either because of increased fertility of members, because beliefs would have led to greater wealth and hence larger populations, or because others would have imitated the more successful. (Boyd and Richerson (1985) discuss imitation of the successful as a method of spreading a cultural trait.) Roes (1995) presents evidence that societies believing in high gods supporting human morality are larger than others. His argument is that larger societies have greater value of gods to enforce agreements, but the converse is more likely: societies with high gods that enforce morality are more likely to be large.

Rules Governing Individuals: The Instability of Libertarianism

Many religions regulate aspects of individual behavior, although Boyer (2001) argues that the moral beliefs precede religion and that religion free rides on certain innate moral tendencies of individuals. It is possible that the religion both free rides on other moral tendencies and strengthens these tendencies by convincing some (who might otherwise cheat) such cheating will bring punishment.

But whatever the source, these beliefs are inconsistent with libertarianism. A default legal regime might be libertarian—a regime that interfered minimally with individual freedom. Nozick (1974) is a modern advocate of such a government. Libertarian beliefs appeal in a normative sense to many economists (including the author). Analytically, libertarianism is a useful starting point. In a libertarian regime, government would define and protect property rights, enforce contracts, and provide true public goods, but would do nothing else. A purely selfish rational person would desire exactly such a regime. Because all legal systems go beyond this level and because citizens generally prefer that government does go beyond the minimal level, there is something of a puzzle. As a positive matter, it appears that there have been few if any fully libertarian governments.

The argument here is that libertarianism as a strategy in the EEA would not have been viable. This should not be read as implying that formal governments existed in the EEA. The point is that selection would have been for preferences for intervention in the lives of others essentially because in a group-living species, actions of individuals create externalities for other members of the group. Apparently, we have evolved preferences to control these actions. These preferences today lead to advocacy of certain laws; in the EEA they might have led to norms associated with interference in the behavior of others.

As a result, individuals with libertarian preferences (who would have established a libertarian order) would have been less successful than individuals with other preferences, so those with libertarian preferences would have left fewer descendants, and such preferences would have been selected against. In other words, establishing a libertarian set of rules would have not been an evolutionarily stable strategy (ESS) for members of a group, and individuals with preferences for such a society would have been reproductively unsuccessful and would have left relatively few descendants. Such preferences, however, would not have been completely eliminated but would explain why few today desire such a government.

A libertarian regime would be subject to attack from both within and from without with respect to the group. Surviving humans would, therefore, have developed defenses to such attacks. These defenses would have taken the form of preferences that made libertarianism difficult or impossible. What is interesting is that these defensive behaviors and preferences generate behaviors that still exist. I discuss them with respect to each source of attack. I then discuss the issue of whether libertarianism would be viable under today's conditions.

Instability of Libertarianism: Attack from within the Group

Under libertarian institutions, some males would have been more successful than others. An individual maximizes fitness by having the largest possible number of mates, because this increases reproductive potential for a man. Most societies have been polygynous and some men, despots in decidedly nonlibertarian societies, have fathered amazingly large numbers of children (Betzig 1986). Though the number of males and females is approximately equal, polygyny for some means sexual abstinence and lack of reproductive success for others, a situation that generates social conflict. The most successful modern societies have been those that outlawed polygyny (at least in its most extreme forms) and thus reduced the internal level of conflict, as discussed in chapter 5. Religion may have played a part in establishing monogamy (see Alexander 1987, 71; MacDonald 1995).

In a world of polygyny, the more successful males would have had a disproportionate number of wives, and other males would, therefore, have left no genes in the surviving gene pool. In this sense, libertarianism could be invaded from within: Those favoring a libertarian society would have allowed some males to become much more successful and acquire a disproportionate share of females. This would have reduced the number of genes of the tolerant (libertarian) individuals in future generations, so the genes of nonlibertarian individuals would have increased in the pool. Of course, the successful males might have been libertarians as well. But they might not; the survival of their genes would not depend on their political beliefs. Unsuccessful libertarians, however, would have done worse than unsuccessful egalitarians, and the number of libertarians would have been reduced over time.

The counter to this form of libertarianism would have been some sort of egalitarianism that is common among groups studied by anthropologists, as discussed in chapters 3 and 5. Members of egalitarian groups do not allow individuals to become excessively dominant. An effect of this behavior is that the otherwise dominant individuals will not be able to engross excessively many females, and the genes of the nondominants would be able to survive. We are descended from individuals who behaved in this way. Boehm (1999) argues that the long period in which egalitarianism was a basic human behavior would have had an effect on our genetic makeup. An alternative might have been to outlaw polygyny. However, in the EEA (and even today, in societies that have engaged in major wars) this would have meant that some fertile women would not mate and have offspring, thus reducing the size of the group and its ability to defend itself. One input in creating this egalitarianism is a lack of privacy, discussed by Sober and Wilson (1998, 176-177). Privacy is also the focus of Posner (1980), Hirshleifer (1980), and Epstein (1980). Egalitarianism for the purpose of reducing the wealth of the most successful must be distinguished from egalitarianism for the purpose of redistribution to the poor, as discussed in chapter 3.

In modern society, egalitarian behavior persists in several ways. Most directly, most modern societies forbid polygyny. This reduces the ability of dominant males to acquire additional mates at the expense of less dominant males. In addition to monogamy, Alexander (1987) lists the graduated income tax, women's suffrage, affirmative action for minorities, and assistance to the handicapped as modern policies aimed at maintaining social cohesion. Sober and Wilson (1998) indicate that policies aimed at maintaining group unity have been important in human evolution and that virtually all societies have intervened in nonlibertarian ways in human behavior.

Another strategy remaining from the past is direct egalitarianism. Important political pressures in contemporary society seem ideologically motivated toward government interference and redistribution, even when such redistribution is inefficient and expensive. In a world survey, 54 percent of respondents agree that wealth can grow, but the rest believe that in some degree wealth accumulation is at the expense of others (Inglehart, Basañex, and Moreno 1998). Of course, part of the motivation for redistribution is altruism, as discussed in chapter 3. But this cannot be the entire answer. Some policies (e.g., excessive marginal tax rates) harm the rich without generating any economic benefits to others. Roe (1998) presents examples of policies that result in net economic harms for everyone in society. While economists explain some of this pressure in terms of rent seeking, a large ideological component seems to play a part as well (e.g., Kau and Rubin 1979; Poole and Rosenthal 1997). Thus, the tendency of governments to intervene in the economy in seemingly counterproductive and often inexplicable ways may be a response to tastes evolved to reduce the extent of genetic inequality in evolutionary environments.

This is an area of behavior where normal utility or wealth maximization may conflict with fitness maximization. If an individual is highly productive and creates much wealth, social as well as private benefits will be generated; a productive individual will not normally absorb the entire surplus he will create (Posner 1979). Thus, utility or wealth maximization would imply that all will benefit from such increased productivity and should encourage it. However, if the added productivity is used to engross additional females, or if tastes evolved in an environment where this occurred, then in fact others will become less fit, although wealthier. In this sense, fitness and utility maximization conflict. This may explain why many utility functions seem to contain elements of envy, even though envy is counterproductive with respect to consumption or wealth maximization.

Although he dies not consider it in these terms, Boyer (2001) discusses another aspects of nonlibertarian behavior consistent with the arguments here. Humans seem to desire that others group members follow certain beliefs and rituals. Boyer argues that observing such rituals is a signal that one is a true cooperating member of a coalition. In the extreme, if nonfollowers are not punished, this may reduce the perceived value of cooperation to others and increase the amount of defection. Therefore, some will punish those who ignore rituals. Boyer argues that this is one explanation for the violence associated with some militant religious fundamentalists; they are punishing defectors to show that defection is costly and so maintain the coalition.

Instability of Libertarianism: Attack from outside the Group

The other source of instability of libertarianism would have been from outside the group. A libertarian group would have had difficulty in defending itself. Many noneconomic behaviors currently forbidden or intensely regulated would have reduced the ability of groups to defend themselves from attack, and quite possibly tastes evolved to forbid these and similar behaviors because those without such tastes would not have survived. The egalitarian impulses discussed above are today associated with political liberalism, while the noneconomic regulations of consumption are more generally associated with conservatism. It may be that liberals (egalitarians) are selected from the more cooperative or altruistic members of society and conservatives from the less altruistic. This is a proposition that could be tested using experimental methods.

Most or all societies interfere paternalistically in private behavior of individuals in the name of morality. Often this interference is associated with religious beliefs, but not always. Hinde (1999) and Boyer (2001) point out that religions may adopt preexisting moral codes. Some examples: prohibitions or limitations on use of drugs (including alcohol) or control of such use through ritualization; interference with and regulation of marriage relationships, as discussed above and in chapter 5; restrictions dealing with sexual behavior, including homosexual relationships, relationships with prostitutes, masturbation, and pornography; restrictions on abortion; and constraints on gambling. The fact that such limitations and restrictions are so common means either that there is an innate taste for such interference, or that at the least it is quite easy for humans to learn to favor such restrictions on the behavior of others. Bischof (1978, 50) indicates that "rules of moderation, asceticism, abstinence" are universal. Similarly, Campbell (1978, 79-80) suggests that "stinginess, greed, gluttony, envy, self-serving dishonesty, theft, lust, promiscuity, pride, and anger may be universally prescribed." Inglehart et al. 1998 indicate that 84 percent of respondents in forty-three countries believe that use of marijuana or hashish is "never justified"; for prostitution, 61 percent; for homosexuality, 59 percent; and for abortion, 29 percent. For the United States the percentages believing practices are "never justified" are marijuana, 74 percent; prostitution, 61 percent; homosexuality, 57 percent; abortion, 35 percent.

None of these authors suggests why those behaviors that appear to have no effect on persons other than the individual making the choice are viewed negatively by those who are unaffected. Indeed, the discussions do not distinguish between behaviors with and without external effects. The suggestion here is that such behaviors would have weakened the group in competition with other groups.

Economists in general, and particularly those with a libertarian bent (including myself), are often puzzled by such rules. Why is it in society's interest to interfere with private behavior that has no external effects? This entire set of controls is inconsistent with a regime of individual autonomy and is an apparently unjustified interference with free markets. Moreover, the interferences create illegal markets that themselves have negative effects on others. Since these markets are illegal, normal enforcement devices such as police, contracts, and courts are lacking. Therefore, violence is often used where other tools would be available in legitimate markets—an example is the well-known drug-related murder used to enforce a contract or acquire territory. Additionally, the risk of selling in such markets means that prices will be higher than would otherwise be true in order to compensate sellers for risk. Therefore, some buyers may engage in crime to support their illegal purchases. Because quality cannot be assured in these markets, contaminated products—drug overdoses—sometimes harm or even kill purchasers. These characteristics are true of illegal drugs today and of alcohol during Prohibition. (Think of the gang wars over territories and of the dangers of bathtub gin.)

These additional costs make the illegality of these markets particularly puzzling and difficult to understand. Economists distinguish between public goods (goods that are consumed be everyone in society, such as national defense, environmental amenities, and public health) and private goods (goods that effect only the individual consuming them, such as food or clothing). It is generally thought that some government intervention is needed to influence consumption of public goods but not for private goods. But the items discussed above are neither (in any conventional sense). That is, my concern that you do not use drugs is neither a private nor a public good. (Of course, in a trivial sense, we could define living in a drug free society as a public good, but this is stretching the terminology—unless there are religious benefits.) Thus, an entire class of behaviors is excluded from conventional economic analysis.[1]

My contention here is that all of these rules can be explained as enhancing group survival in an environment in which there are struggles and combat between competing groups. In this context, it is important to note that young males are extremely important in defense, because they now and traditionally have made up the warriors of society. Young males are also disproportionate risk takers for evolutionary reasons (Rubin and Paul 1979). That is, if a young male avoids a risk but, as a result, does not find a mate, his genes are eliminated, so this behavior would die out, as it would if he takes a risk and dies. But if a young male takes a risk and succeeds, he is more likely to mate and leave descendants. Therefore, young males are risk takers, and they would be expected to engage in the forbidden behaviors if allowed.

Young males may compete for status and the associated fitness gains. This is generally called sexual competition (Miller 2000). Many behaviors of young males may be understood as a form of such competition. Some—sports activity, for example—are harmless. Others—say, aggression against neighboring tribes—may be harmful overall but beneficial to the group. Miller claims that much human activity (including, e.g., much of the arts) is a form of such sexual competition. Indeed, Miller claims that such competition is one of the most important human activities and has driven much human evolution. We need not decide if Miller is correct, but clearly some activities do fit his model. The competition generally takes the form of demonstrating strength, intelligence, or other signs of fitness.

One form of such competition discussed by Miller following Zahavi and Zahavi 1997 is handicap competition. This is competition through demonstrating that even when suffering from a handicap (which may be self-imposed), an individual is still powerful. Gintis (2000a) indicates that handicap competition may be the most common form of sexual competition. Suffering from a handicap is then a sign of fitness, and the argument is that we (and particularly young males) may be selected to engage in such competition. One form of handicap competition may be through consumption of harmful substances. It has often been considered a mark if manhood to be able to hold your liquor and similar attitudes may apply to illegal drugs. (Diamond 1992, makes this point as well.) Such competition may be individually rational in that those who are more successful may also be more fit. But engaging in this form of competition is clearly a prisoner's dilemma. If all young males could agree not to engage in this form of competition, they would all be better off (if they could find less destructive forms of competition), but any one male may benefit from the competition. Evidence indicates that males are much more likely to use alcohol, marijuana, cocaine, and heroin. Moreover, use is more common among those between twelve and thirty years old, and decreases as marriage increases (all results from Saffer and Chaloupa 1999). These observations are all consistent young unmarried males using these substances as a form of handicap competition. Social prohibitions on this behavior, backed up by religion, may be a socially useful way of limiting such harmful competition.

To the extent that consumption of harmful substances such as drugs is a form of sexual competition, we may devise some useful strategies for reducing such consumption. Government sponsored advertising aimed at discouraging use of these substances stresses the harm that they may cause. If consumption is, however, a form of handicap competition, then these ads may be counterproductive. Handicap competition is only useful if the consumed substance has real costs, and emphasizing these costs may, therefore, increase the value of the substance in this form of competition. A more useful message would be that only the unsuccessful must resort to this form of competition. If advertising and public service messages could convince potential users that only losers use drugs, the value of these substances as signals of strength would be reduced.

Many of the puzzling social regulations may be viewed as controlling the behavior of young males so as to preserve them as potential fighters. For example, restrictions on drugs and alcohol mean that those who fight will generally be more effective at it. To the extent that gambling might lead to reduced incomes and thus physical weakness for some (in a subsistence society), again it would reduce the fighting strength of the society.

The common restrictions on sexual behavior and on abortion will sometimes have led in the EEA to increased population and thus more potential warriors. Hrdy (1999) explains opposition to abortion as part of an evolutionarily old desire by dominant males to control fertility of females. This is not inconsistent with a desire for larger population sizes, since this desire apparently extends to females who are not carrying offspring of the males themselves. In other words, the size of the population in a society is a public good (to use the economists' term) in the sense that everyone in a society must share the same population. That is, if I live in a tribe of two hundred persons and you are also a member, then you also live in a tribe of two hundred persons. If the size of the tribe is relevant for survival, then each of us may have a desire to control that size or to control the behavior that influences population size. In the EEA, the important determinants of population size had to do with fertility-related behavior, and so we may have tendencies to want to control that type of behavior. But whatever the evolutionary basis for the taste against abortion may be, this preference is counterproductive today. We do not need a larger population. If we did, in the United States at least, we could increase our population as much as we wanted simply by allowing increased immigration. That is, population can be controlled with more effective tools than were available in the EEA. If the basis is simply a desire of some individuals to control behavior of others (the feminist position, expressed in biological terms by Hrdy), this control is inconsistent with a free society. Moreover, limiting of abortion would lead to unwanted births. Evidence shows that at least some of these unwanted births lead to increased social costs, such as increased crime (Donohue and Levitt 2001).

Many of these prohibitions and restrictions are associated with religion and a belief in the supernatural. The same argument that explains the moral restrictions can also explain religion. If a society adopting such restrictions is more powerful and successful than one that does not, mechanisms to increase the probability of such adoption will be selected for. If an individual has no rational private reason to refrain from certain behaviors and potential gains in utility from engaging in them, a society or tribe (or species) in which individuals can be led to believe in a supernatural benefit from avoiding undesirable behavior will thrive relative to one lacking such a mechanism.

In other words, religion can serve to solve the prisoner's dilemma: Don't drug yourself or drink yourself senseless, or abort your child, or else you will go to hell, a private cost. This belief will avoid the social cost of losing the next battle to the guys in the next valley who do have such prohibitions and who will otherwise exterminate us. A society with beliefs and taboos of this sort could defeat one without them, so there is exactly selection pressure for such beliefs and taboos. Cooperation through avoiding behavior that weakens the group becomes a rational strategy if one believes that defection will have long (very: eternally long) run costs. Burkert (1996) discusses religion in similar terms, although he does not discuss the role of religion in solving prisoner's dilemmas.

On the other hand, for any individual, not sharing such beliefs can yield a payoff. This is one of the prisoner's dilemmas mentioned earlier. However, several authorities have recently shown that if individuals in a society are willing to punish noncooperators and also to punish those who fail to punish noncooperators, then cooperation can evolve in a society, although inefficient behaviors can also evolve under these circumstances (Hirshleifer and Martinez-Coll 1988; Boyd and Richerson 1992; Axelrod 1997). These models can explain human willingness to obey rules. If some individuals believe in the tenets of a religion and are willing to punish nonbelievers, then all can be made to behave as if they are believers. Sober and Wilson (1998) indicate that these rules of punishment have traditionally been used to enforce group cohesiveness. They indicate that enforcing such rules has traditionally been relatively low cost, as social ostracism has been available as an enforcement device. In today's more anonymous society, enforcement may have much greater costs. I now discuss this issue.

Is Libertarianism Viable Today?

The argument to this point can explain why most people living today have nonlibertarian preferences. However, these tastes presumably evolved in the EEA, and it is possible that conditions have changed sufficiently so that a libertarian society would be more viable today than would have been true under evolutionary conditions. I now explore that possibility. It appears that the benefits of interventionist preferences may have decreased and the costs of enforcing these preferences may have increased, so that, even if these preferences were fitness maximizing in the EEA, they may no longer be so today.

BENEFITS

First consider defense. Would others conquer a libertarian society under today's conditions? In the Vietnam War, it was widely reported that many soldiers used drugs and the United States lost that war, indicating that such behavior might still incur costs. (However, it is generally claimed that drugs were not used during combat: see, e.g., Appy 1993, 283–284.) On the other hand, the Gulf War and the Kosovo war were much more capital intensive (perhaps because of the lack of a military draft and the corresponding inefficiently low price of personnel) and were fought largely through advanced technology. Thus, combat behavior of individual soldiers was relatively less important, and this may be true of future United States wars as well. Increasing capital intensity of warfare suggests that the military basis for some nonlibertarian preferences no longer applies, at least for the United States. Browne (2001), however, argues that the Gulf War was atypical and that human strength and endurance will be important in future wars as well, since infantry will generally be needed to capture and hold territory. He shows that the claim that war has changed and that physical strength is now unimportant has been made since at least World War II, and has generally been incorrect. If so (and the case he makes is persuasive), then nonlibertarian preferences leading to increased strength of young males will still be important.

The United States is large enough that in foreseeable future wars most young males might not be needed to fight. We could select those who were needed as members of a paid army, and nonuse of drugs could be a criterion for such selection. The cost might be a more expensive army, since this policy might exclude some potential enlistees, but this cost would probably be less than the cost of our current drug policy. For other smaller countries fighting, for example, border wars with neighbors or engaged in the ethnic conflicts discussed in chapter 2, manpower may still be an important factor for military success, and such countries might find restrictions on behavior desirable to increase the number and effectiveness of their soldiers. Policies aimed at increasing population such as prohibitions on abortion serve no purpose in the United States because we could use immigration policy to increase our population to whatever level we might desire.

Thus, those restrictions on behavior that were apparently aimed at creating military power or at increasing the fighting ability of young men might serve no useful purpose in today's world, at least in the United States. Costs of satisfying nonlibertarian preferences are not trivial or peripheral; significant amounts of resources are spent on them, and some of these policies have large social costs in terms of disruption. Abortion is a major political issue in America today, even though most of the participants in the debate would not likely be directly affected themselves. Those opposed can simply avoid having an abortion. Many of the strong proponents are relatively wealthy and would be able to obtain an abortion for themselves or their female relatives even if it were made more difficult. Nonetheless, the debate over this issue is socially costly. Other examples of apparently strong moralistic preferences would be the effort devoted by many to establish prohibition of consumption of alcohol and on the war on drugs. Large amounts of wealth and of political capital have been spent on such issues. In addition to the direct cost of these efforts, large indirect costs in terms of the additional crime generated by the policies have been incurred.

Another benefit of nonlibertarian preferences discussed above is the possibility of increased social stability, which would have been associated with survival in a competitive environment such as the EEA. Two interventionist policies generate such stability. First is outlawing polygyny. Outright polygyny would still probably generate sufficient social tensions so that outlawing this behavior is probably still useful. On the other hand, the current set of divorce laws does not seem to create any excess social tensions, as discussed in chapter 5. The second nonlibertarian policy aimed at creating stability is limiting high incomes, in an environment where high and unequal incomes would have lead to reduced fertility of less successful males. This preference is associated with envy, as discussed in chapter 4. However, because polygyny is not a problem or an issue today, this policy should also serve no function. Moreover, the policy is more costly now than in the EEA. In the EEA, creating wealth was difficult, and the major social benefit of wealth was for use as social insurance (Posner 1980). The major way to accumulate wealth in such a society would be through hoarding and refusal to participate in social exchange, particularly in a hunter-gather environment where storing wealth would be difficult. On the other hand, today, most wealth creation is associated with creation of surpluses for others, so limiting wealth because of envy would have much larger costs today than in the EEA.

COSTS

The argument so far is that benefits of nonlibertarian tastes may be lower today than in the EEA, where such tastes evolved. On the other hand, costs may be higher. Enforcement of antilibertarian social norms may generally be more expensive today than was true in the EEA. Today's societies are much larger and more anonymous than the relatively small, homogenous societies of the past. As indicated above, Sober and Wilson (1998) believe that enforcement of norms was relatively inexpensive in the past, where information about behavior was available at low cost (because of lack of privacy) and enforcement (through, e.g., banishment) was also inexpensive.

Today, enforcement is much more expensive. Information about behaviors is not easily or cheaply available because of increased mobility and privacy and the anonymity associated with large-scale societies, and enforcement requires relatively expensive incarceration rather than inexpensive banishment or ostracism. Therefore, even if the evolution of nonlibertarian tastes in the EEA can be rationally explained, some such tastes are counterproductive today, even if benefits are the same. Moreover, as discussed above, to the extent that nonlibertarian preferences lead to divisiveness in our large, heterogeneous society, then the benefits of these preferences may be reduced.

These questions seem to be worth further research. At a minimum, however, the argument provides a potential explanation for social prohibitions that otherwise are puzzling to many. Even if one favors libertarianism, it would be useful to understand why most individuals do not in order to determine if a movement in that direction would be useful or feasible. In proceeding, it is important to note that individuals might have preferences that evolved for some reason without necessarily understanding or accepting that reason.

Summary

Even though religious belief is nontestable, many people still have such beliefs and are willing to spend substantial resources to advance their beliefs. The best explanation for these beliefs is anthropomorphism. Humans attribute intentionality to even inanimate causes, and religion is an extreme example of this anthropomorphism. The form of religion is not determined by such factors, and since religious beliefs are nontestable, the form is indeterminate. However, selection pressures have led to certain characteristics of the modern large inclusive religions—Christianity and Islam. Both religions are nonethnic, in that they actively seek converts from all ethnic groups.

Moreover, both attempt to regulate private behavior in various ways. Some of this regulation is to solve various prisoner's dilemmas, as when cheating and breaking of contracts is forbidden. Some prohibitions are more puzzling. These include prohibitions on various private behaviors that have no obvious impact on the welfare of others, such as regulation of drugs and alcohol, of sexual behavior, and of abortion. I explain these regulations by arguing that a libertarian regime (one that did not regulate such behaviors) would not have been viable in the EEA. Such a regime would have been subject to attack from within, as more powerful males engrossed excessive numbers of women. A libertarian regime would also have been subject to attack from outside, as in a libertarian regime young males—the basic human fighting force—would have had incentives to engage in behavior that might have made them less suitable as fighters. Therefore, there would have been incentives for interventionist preferences to evolve.

Whether these preferences still serve a useful function today is an open question. Benefits of intervention may be smaller and costs larger than in the EEA. Policies aimed at increasing population, such as antiabortion policies, serve no purpose in the United States because immigration policy is a better way to influence population size. The theory of sexual selection and handicap competition may provide a useful way of discouraging consumption of harmful substances through emphasizing that only the unsuccessful would use such substances. On the other hand, current ads stressing the harm of these substances would increase their value as signals.

Religions also serve useful purposes. By inducing cooperative behavior and threatening punishment for noncooperative (in this life or the next) religions can solve many prisoner's dilemmas and increase productivity by strengthening cooperation.

CHAPTER 7

How Humans Make Political Decisions

One theme of this book has been the contrast between the small societies of the EEA and the incomparably larger societies in which we now live. In evolutionary times, societies were sufficiently small that everyone knew everyone else, at least by sight. Human bands probably contained thirty to fifty individuals with whom each person interacted on a regular basis. Larger groups of about three hundred to five hundred probably came into being when smaller bands would meet for various purposes, such as seeking mates (Caporael and Baron 1997). Now we live in societies of up to one billion persons (India, China). While this difference has effected many of the issues discussed in this book, it is perhaps nowhere more important than in discussing ways of making political decisions. In societies in evolutionary times, our ancestors knew everyone who would be involved in a decision; today we effectively know no one. Our institutions have had to adapt to this difference and they have done a remarkable job of adapting. But decision making bears many marks of our evolutionary history. This chapter is the story of that history and those adaptations.

I first discuss an issue involving preferences—the difference in risk preference between men and women, and the political ramifications of this difference. The rest of the chapter discusses methods of processing information. I discuss the debate among social scientists as the level of rationality of human decision makers and indicate that evolutionary theory has much to say about this debate. I then analyze some manifestations of information processing, including the role of identified individuals in political decision making, the importance of the status quo, and self-deception. Finally, I discuss the implications for our evolved decision making propensities on jury behavior.

Men, Women, and Risk

One theme that I have stressed is the difference between men and women in political decision making. The benefits of political behavior in evolutionary terms are measured in terms of increased fitness. Males exhibit a greater variance in fitness than females, which means that the potential gains from political success are greater for males. As a result, politics has been more important for males than for females. This shows itself in many ways. For example, in chapter 5, I indicated that men are generally more politically active than are women.

This same difference is reflected in attitudes about risk. Think of a person in the EEA who has sufficient income to survive and raise offspring. If that person is a female, increased wealth or power will have a relatively small impact, since she will have as many children as biologically possible anyway. Survival of children, however, has always been an issue for women, and so women will be more concerned with obtaining enough resources for their existing children. On the other hand, men will be more concerned with obtaining additional wives (Hrdy 1999). Increased wealth may have some impact for females in that it may lead to more biological success for her children. But this will be discounted by both relatedness and by time, because it will not happen until the children are grown. Nonetheless, in societies where women have been politically active many of the benefits have gone to their sons (Low 2000).

The situation is different for a male. In a polygynous society—which most societies were in the EEA—additional wealth or status can translate into additional mates and, therefore, additional fitness. Males would have more to gain from increases in wealth than would females. That is, men are concerned with number of children, while women are more concerned with quality. If faced with some gamble that might either increase or decrease wealth, males would have a greater incentive to accept that gamble, since it might translate into increased fitness through an increase in the number of mates. Interestingly today, at least in the United States, whenever a man quits some major public activity (either in government or in business), he indicates that he wants to spend more time with his family. In the evolutionary environment, this might have been a signal that he was planning activities that would maximize the fitness of his existing family, rather than seeking to acquire additional mates.

Females would have little incentive to gamble, since they would have much less to gain if the gamble paid off unless their incomes are so low that the status quo without the gamble would lead to death of the woman or of her living children. This should imply that women in the political arena are more risk averse than are men. This is indeed what we find. For example, consider income taxes and transfer payments. Income taxes reduce the gains to higher earnings because some of these earnings are taxed away. Transfers increase the incomes of low earners. Thus, these policies reduce the spread of the income distributions and so reduce risk. A risk-avoiding person would be more in favor of these policies than would a risk-seeking person. The risk seeker is willing to give up some of the certainty associated with transfer payments in return for a greater chance at the high rewards associated with low taxes at the top. We would, therefore, predict that women as voters would be more in favor of taxes and transfers than would be men. This is exactly what we find (Lott and Kenny 1999). As women have been given the right to vote, the size of government and the amount of redistribution have increased. This is consistent with greater risk aversion on the part of women. Additionally, women judge risks of various technologies as being greater than do men, again in a manner consistent with differential risk aversion (Slovic 2000; however, Slovic notes that this differs by race, with white males having relatively low fears of technologies.)

We would also expect women to be less interested in conflict. War in evolutionary times was a way for men to get more women; it had little advantage for women, whose fitness was limited by their own reproductive potential. Women who already had offspring and who were put at risk by war would gain little from a victory. A woman who was on the losing side of a war might have expected to be captured by the enemy and made a mate of one of the victors, so that from a biological perspective the outcome of the war or conflict would have been extremely unpleasant and probably fitness reducing relative to remaining in her original home. This may be why women are known as being more in favor of peace than are men. (For a discussion of the difference in military preferences between men and women, see Browne 2001.) It might appear that the world would be better if women had more power, so that the amount of war and conflict would be reduced. This may be true—but it is another prisoner's dilemma. If some societies give additional political power to women and these societies become less warlike, they are in danger of being attacked and conquered by other societies that have not made this decision.

It is important to stress that asking which set of preferences is correct or better is meaningless. Preferences are the things we use to measure better or worse, and so the terms cannot be used to characterize preferences themselves. Different preferences will lead to different outcomes in the political marketplace. For example, when women were allowed to vote, the nature of political outcomes changed. But it is not meaningful to ask which set of preferences is better. We can say that outcomes are more representative if women vote than otherwise and to that extent they are better.

Rationality

A major debate in the social sciences—and primarily in economics and psychology—is the extent to which humans make decisions rationally. The state of the debate as I write this book is as follows.

Economists always assumed rationality. That is, the basic assumption made by economists is that all actors in the economic system (consumers, investors, producers) were completely rational. As used by economists, rationality meant that rational methods were used to achieve ends; ends or goals themselves were considered as being arbitrary, and neither rational or irrational. To say that ends were well adapted to achieve goals meant that behavior followed certain axioms. One of the most important was consistency; if A is preferred to B, and B is preferred to C, then A is preferred to C.[1]

Cognitive psychologists—primarily Daniel Kahneman and Amos Tversky and their students (e.g., Kahneman and Tversky 1979; Kahneman, Slovic, Tversky 1982)—challenged this assumption. They produced substantial amounts of experimental evidence showing that humans in many circumstances did not make rational decisions. Subjects committed many different kinds of errors, generally leading to some inconsistency that violated the standard economic model of rational decision making. Although economists resisted this finding for a long time, there is now some agreement among many economists that in fact humans are not so rational after all (see e.g., Rabin 1998 and Thaler 1992).

There is a third set of players, besides economists and cognitive psychologists. These are the evolutionary psychologists, the scholars whose outlook is most consistent with the approach of this book. From an evolutionary perspective, it is difficult to understand why much human decision making would be irrational. First, the brain is a biologically very expensive organ (consuming about 20 percent of energy intake), and humans have evolved a large brain. We would not expect such a large and expensive organ to function poorly. Second, humans are deeply competitive with each other, as discussed in chapter 2. In this competition, those who made better decisions would have done better. Again, this leads us to expect that humans are not likely to be particularly irrational.

In fact, evidence indicates that we actually do a better job of decision making than the cognitive psychologists evidence would indicate. The problem is that in many cases the experiments that purport to show that decisions are made incorrectly do not fully consider the evolutionary background in which our decision-making capacity evolved. If problems are posed in ways consistent with our evolved abilities, subjects are much more likely to get the answers right. The psychologist who has done the most to show that decisions are not as incorrect as might appear is Gerd Gigerenzer (see, e.g., Gigerenzer 1991). Much of his research has dealt with probabilities. Cognitive psychologists (and many experimental economists) often ask subjects questions about probabilities; in certain circumstances the subjects reliably get incorrect answers. However, Gigerenzer (and others, including Cosmides and Tooby 1996), reasoned that in normal human environments, and in evolutionary environments, we do not routinely calculate probabilities. But we do observe frequencies of events and make judgments based on these frequencies. Thus, Gigerenzer has recast many problems involving probability estimates as problems involving frequencies, and subjects do much better under the latter condition.[2]

I should note that there is a debate between Gigerenzer (1996) and Kahneman and Tversky (1996). I do not want to interpret this debate; anyone interested can read the authors themselves, and the topic is somewhat peripheral to this volume. I do believe, however, that in the long run Gigerenzer and his associates will prevail. This is because his school uses explicitly evolutionary reasoning in the analysis, while Kahneman and Tversky seem hostile to this form of argument. For example, Kahneman and Tversky (1996, footnote 3) refer to evolutionary reasoning (by Cosmides and Tooby 1996) as "speculative." Of course, if I did not believe in the power of evolutionary reasoning, I would not be writing this book.

I now discuss some examples of alleged irrationalities in decision making and consider their evolutionary basis.

Overconfidence Bias

One example is overconfidence bias. Ask subjects a series of questions whose answers they are unlikely to know (e.g., ask Americans whether Hyderabad or Islamabad is larger in population). Then ask how likely it is that their answer is correct. Subjects generally assert that the probability that they are correct is larger than in fact turns out to be true. They overestimate the probability that they are correct; this is the overconfidence bias. Gigerenzer then restated the problem in terms of relative frequencies. Rather than asking how confident a subject was about each answer, the problem posed was, after a series of questions, What fraction of the questions do you think you got right? Although from a mathematical perspective this might be the same problem, it turns out that subjects do much better on the latter question, and the overconfidence bias disappears.

Conjunction Fallacy

Another alleged bias is the conjunction fallacy. This is a fallacy of indicating that the conjunction of two events is more likely than one event. The classic example is the Linda story:

> Linda is 31 years old, outspoken, and very bright. She majored in philosophy. As a student, she was deeply concerned with issues of discrimination and social justice and also participated in anti-nuclear demonstrations.

Subjects are then asked about probability of various situations for Linda, and most agree that it is more probable that she is a bank teller and active in the feminist movement than that she is a bank teller. It is impossible for a conjunction of two events to have a higher probability than a single event, so the majority answer is clearly wrong. But again, if the question is asked in frequency terms (of 100 persons like Linda, how many are bank tellers? bank tellers and active in the feminist movement?) subjects are much more likely to give a correct answer.

Base Rate Fallacy

The third example is the base rate fallacy. A classic example used to illustrate the fallacy:

> If a test to detect a disease whose prevalence is 1/1000 has a false positive rate of 5 percent, what is the chance that a person found to have a positive result actually has the disease, assuming you know nothing about the person's symptoms or signs?

Here, the correct answer is 2 percent, and the typical answer (from students and staff at Harvard Medical School!) is 95 percent. But again, in careful replications using frequency terms rather than probabilities, Cosmides and Tooby (1996) have gotten correct answers from between 76 and 92 percent of subjects.

Hindsight Bias

Gigerenzer et al. (1999) examine hindsight bias. Hindsight bias is found when someone is asked to estimate the probability of some event before it occurs and then asked later what the prior estimate was. It is commonly found that people ex post indicate that their ex ante estimate was higher than it actually was for events that did occur. For example, if we were to ask people in 1999 what are the chances of a stock market fall in 2000, they might say 50 percent. Should the market fall and the same people be asked what their estimate was before the fall, they would give a higher number. This is explained as a natural result of two processes. When asked for the ex ante estimate, most people cannot recall it, and so go through the same calculation process that was used to provide the original estimate. However, because we routinely update information as new data becomes available, when we perform the calculation, we get a bias because we automatically use the new data in the analysis. The costs of any hindsight bias are less than would be the costs of increased memory capacity to avoid this problem. In a political context, hindsight bias means that if a politician makes an incorrect decision, voters will blame him for it to a greater extent than is justified because of hindsight bias—any fool should have seen that that was a dumb move. This will lead politicians to be excessively careful in avoiding decisions that might turn out wrong and could lead to excess caution in decision making.

Two Decision-Making Mechanisms

Evans and Over (1996) provide a useful theory of the extent of rationality. They argue that there are two separate decision-making mechanisms, which they call Rationality$_1$ and Rationality$_2$.[3] Rationality$_1$ involves thinking and acting "in a way that is generally reliable and efficient for achieving one's goals," while Rationality$_2$ involves thinking and acting in a way that "is sanctioned by a normative theory" (8). In other words, Rationality$_1$ is an efficient way to achieve a purpose, though it may not meet the standard criteria for rational behavior. The point is that in actual behavior we use information and shortcuts that may not be strictly logical but that are efficient in the real world. Basically, this involves making use of the structure of the real world, which is deep and complex and difficult to model in a formal way.

Moreover, we generally do not understand or verbalize the methods of making decisions in Rationality$_1$, although Rationality$_2$ methods are self-consciously discussed and explained. Some of the experimental evidence cited above can be explained if subjects were using Rationality$_1$ when experimenters wanted them to use Rationality$_2$. For example, in the hindsight bias experiments, subjects are told to use information available in the past, but they actually use the new information about the actual outcome of an event without necessarily knowing that they are doing so. Evans and Over (1996) discuss belief bias. Subjects who are asked to evaluate the logic of arguments are more likely to find the argument correct if they know that the conclusion is true. This is an error in the context of the Rationality$_2$ test of logic, but a correct inference in the Rationality$_1$ world where we make use of all relevant information.

The argument that decision making makes use of the structure of the world is similar to that in Gigerenzer et al. (1999). The underlying argument of the book is that the environments in which we evolved and in which we now live have certain regularities and that decision-making mechanisms—both evolved mechanisms and the mechanisms that we actually use today—take advantage of these environmental regularities. The models economists use and cognitive psychologists, who assume generalized all purpose decision-making mechanisms based on pure logic, criticize are not the models that humans actually use in decision making, which is ultimately why experimental evidence rejects some of these models. However, the mechanisms that we actually use may be close to optimal in the environments where we use them, so the pessimistic conclusions of those opposed to rationality may not be correct either. In other words, we have evolved to achieve a certain level of rationality, and more rationality than this would not pay in the environments where we actually function. For example, the computational cost of more complete analyses in many circumstances would not be worthwhile.

Most of the Gigerenzer et al. (1999) book illustrates this argument by showing that in many circumstances shortcut decision-making mechanisms (the simple heuristics of the title) are remarkably accurate. This is essentially because the simple heuristics are able to take advantage of features of the natural environment that simplify decision making. Moreover, this is the link to evolution. We have evolved in natural environments, and we have evolved to use mechanisms that benefit from the structure of these environments.

The book begins with a discussion of "Fast and Frugal Heuristics." These heuristics are fast because they use only part of the potential available information in the environment, and they are frugal for the same reason. Examination of these heuristics is part of Herbert Simon's bounded rationality research program (e.g., H. Simon 1997). They are contrasted with various demons—agents with unlimited reasoning ability that always make correct decisions. There are three parts to a fast and frugal heuristic: guiding search for information or alternatives, stopping search, and making a decision. There are particular heuristics for particular problems; decision makers choose the correct heuristic for a problem. Different environments require different heuristics, and this tradeoff of specificity for generality is part of the reason fast heuristics work as well as they do.

Much of the rest of the book demonstrates that these heuristics do indeed work remarkably well. The first one considered is the recognition heuristic—a process based on assuming that something we have heard of is higher ranked than something of which we are ignorant. An example from the book: cities whose names are familiar are likely to be larger than cities with unfamiliar names. In some environments and for some problems, more ignorant individuals can outperform those who are more informed.

The recognition heuristic could easily have evolved as an important human decision-making tool. Some examples: A person whom one recognizes is not likely to be an enemy. A recognized food is unlikely to be poisonous. After a hard day hunting or gathering, search for recognized landmarks can lead one home. In modern economies, it may be that much advertising is aimed at taking advantage of this heuristic; this provides a simpler explanation for the importance of such advertising than, for example, the well-known Klein-Leffler (1981) mechanism. It is well known that in politics name recognition is quite important; this is an example of the same process.

The next several chapters of Gigerenzer et al. (1999) discuss "One-Reason Decision Making," a method of decision making that uses only one piece of information. Several variants are discussed; the one most frequently used is called "Take the Best," a decision technique where the single piece of information that is judged best is used alone for decision making. This decision criterion is compared with others, including Bayesian decision making and multiple regression; many of the comparisons are made using real world data that was originally used in statistics textbooks to illustrate regression techniques. One conclusion is that these simple techniques can often obtain results as good as or better than those obtained from more sophisticated statistical techniques, such as regression analysis or Bayesian mechanisms.

Political Implications

Actors in the political system might use either the Rationality$_1$ or the Rationality$_2$ method of decision making. Economists, political scientists, policy analysts and (hopefully) bureaucratic decision makers would be expected to undertake explicit calculations and use Rationality$_2$. That is, these agents would be expected to use data and statistical and other formal methods to analyze problems. On the other hand, voters would probably use Rationality$_1$. This is because there is no reason for voters to spend substantial effort and resources calculating probabilities and reaching optimal decisions. Indeed, given the vanishingly small probability that a single vote will influence the outcome of an election, there is no reason for people to vote at all.

This difference between methods of decision making used by voters and by analysts should lead to differences in preferred policies. We find, for example, that experts are concerned with different environmental hazards than are citizens, and this often leads to biased decision making by agencies responding to citizens concerns rather than true risks (Slovic 2000). The concerns of citizens are based on Rationality$_1$ thinking, while those of experts are based on Rationality$_2$. Caplan (2001) cites evidence that non-economists think in a systematically different way than economists. Other implications of the difference in ways of thinking are illustrated below, when specific decision-making techniques are discussed.

It is also true that political decisions may be made with less rationality than other decisions. There are several reasons for this. First, any one voter or citizen has a very small impact on the final decision. Indeed, any one voter has a trivial impact, since the chance of any voter being decisive (i.e., making the final difference in the outcome of an election) is vanishingly small. Second, as a result, voters have little incentive to understand the issues; that is, voters are said to be rationally ignorant in making decisions. Moreover, because individual voters have little impact on the actual outcome, there is no feedback process and so no method of learning that decisions were incorrect. Thus, it may be that voters exhibit many of the cognitive biases and illusions in the political process, even if these biases are not so common in economic decision making. For example, in buying a car for myself, my own decision is determinative—I will get the kind of car I want, and so it pays to study the choice carefully. But in voting, there is only a fifty-fifty chance (in the United States, a two-party system) that my candidate will be elected, and my decision has no impact on this outcome. Therefore, a rational citizen will pay much more attention to deciding what kind of car or breakfast cereal he wants than to who he prefers as president.

Identifiable Individuals

One example of Rationality$_1$ as used by individuals is the attention paid in the political process to identifiable individuals. It is commonly observed that political decision making does base many policies on occurrences involving such identified individuals. For example, we may spend large amounts saving the life of one identified person when the same amount could save the lives of many more unidentified individuals. Posner (1999) points out that charities use pictures of starving children in fund raising, rather than providing statistics on poverty in particular countries. For a discussion of some implications, see Jenni and Loewenstein (1997). (Their paper explicitly omits any evolutionary arguments. For a paper that does include an evolutionary analysis, see Moore 1996.) Other implications of this method of decision making are discussed below.

In the evolutionary environment, humans or their prehuman ancestors lived in relatively small groups of closely related individuals who would have known each other as individuals. In a hunting-gathering economy, there would have been ample scope for fitness enhancing income transfers, as discussed in chapter 3. For example, if one individual had a successful hunt, then there might have been more food than he and his immediate family could consume before spoilage. In this circumstance, transfers would have benefited recipients more than they would have harmed donors, providing incentives for transfers. Moreover, in a society where storage of wealth was difficult or impossible, there would have been few incentives for accumulation, again increasing the benefits of transfers under certain circumstances. In such an environment, there would have been fitness increasing incentives for charity or contributions to welfare of others.

Members of the group with whom an individual came into contact would generally have been relatives, so that any transfer increasing the fitness of the recipient would have been selected for by kin selection. Moreover, since individuals would have known each other personally, reciprocal altruism (Trivers 1971) would have been relevant. Note that any individual with whom one was familiar would have likely been a relative; there would have been no need to distinguish familiar strangers from relatives. Both factors mean that there would have been selection pressure for income transfers to known, identifiable individuals. That is, those of our predecessors who transferred resources to those they knew within the local group would have left more genes in the gene pool from which we are now selected.

On the other hand, as Moore (1996) points out, there is no reason to expect that we would have evolved to perform appropriate calculations to maximize fitness for a large amorphous group of unknown individuals. Such groups would not have existed in the EEA, and so there would have been no evolutionary incentive to learn how to maximize for such a group. The result is that we might expect contemporary humans to be adapted to providing benefits to recognizable, identifiable individuals rather than to anonymous or statistical individuals, even if the net benefit of the latter type of transfer is greater than the benefit of transfers to identifiable individuals. Moore has made the argument that this explains the emphasis of modern medicine on patient care rather than on prevention, and this argument seems correct. The implications are much broader, however. There are many ways in which the general principle pervades public decision making.

One implication is that people may have tendencies to overestimate their own impact on political decision making and outcomes (Baumgartner and Leech 1998). In the evolutionary environment, where perhaps fifty individuals made decisions, each individual would have had an impact on the final decision (Boehm 1999). Now, for most decisions, most individuals have absolutely no impact, since there are simply too many decision makers for any one individual to matter. Nonetheless, we may retain the thought patterns of our small group evolutionary environment. Baumgartner and Leech (1998) point out that this may be the basis for many individuals joining interest groups, and they suggest that groups have an incentive to induce members to overestimate their contribution. It may also be part of the explanation for voting in a situation in which any one vote has zero influence. We are simply not suited to understand situations in which our decision has no influence.

I will mention briefly another aspect of the significance of identified individuals. That is the importance of stories for human learning. I hypothesize that this preference evolved because such stories are an efficient method of learning about the environment, where the environment includes other individuals as an important component (Scalise Sugiyama 2001). Before statistical analysis and other aspects of Rationality$_2$, learning what happened to others was a good way to learn what might happen to you. Dunbar (1996) stresses that gossip was one way to learn about such events. But stories may have been another. It appears that the types of stories we like are those that would have been relevant for learning in the EEA. For example, young males like stories of violence and action, and, as mentioned in chapter 2, young males were the warriors. Stories of crime are about social cheating, and we have evolved to learn to be aware of such cheating. In foraging societies, stories discuss behavior of animals in ways useful for learning about finding and killing these animals (Scalise Sugiyama 2001). It may also be true of painting and sculpture and also ways of learning. Since the photograph and the invention of writing, we have more efficient ways of learning, but for preliterate people a picture could provide a substantial amount of information.

Implications

This evolved bias affects many policy decisions. In this section, I discuss some of these decisions.

POLITICAL DECISION MAKING

Congressional hearings and other political forums often feature particular individuals testifying about particular issues that have effected their lives. To public policy analysts or economists, such testimony makes no sense. If there is sufficient information regarding some problem to justify legislative or regulatory action, than the objective, statistical information should be used. Testimony by random individuals should carry no weight, and it is a puzzle as to why decision makers rely on such information.

But the perspective provided here makes it clear that such testimony does serve a useful purpose with respect to achieving passage of a law. Testimony indicates that particular, identifiable individuals are affected by some problem. For voters and perhaps for politicians themselves, identifiable individuals are relevant for decision making, and statistical abstractions have relatively less value. Indeed, voters who are concerned about identifiable individuals will select politicians who share these concerns; potential politicians who make arguments based on statistical abstractions may be relatively less successful than politicians who use identifiable individuals as evidence for positions. Moreover, this method of argumentation applies to both liberals and conservatives; President Reagan, the most successful conservative politician of recent times, used anecdotes and stories of particular individuals to make political points. Indeed, the difficulty economists and others of their ilk have with understanding the nature and purpose of testimony of identifiable individuals in Congress and elsewhere may explain their relative lack of political success. Lawyers, on the other hand, deal with individual cases in their professional lives and, therefore, may be better suited to make the kinds of arguments that voters appreciate and understand.

Identifiable individuals in the political process play another important role. Politicians themselves are identified individuals. In electing politicians to office, we seem to rely to a remarkable extent on looking for characteristics that would have been useful for chiefs in a tribe, but may be less important in a modern politician. For example, taller politicians do relatively better in wining elections. Height is useful for killing enemies in hand-to-hand combat, but less useful for a United States president who is unlikely to get into situations involving this activity. Personality characteristics are probably more important in choosing politicians than is rational. We might want an honest politician, so that we can expect him or her to attempt to fulfill promises. Other characteristics (whether the person is a good spouse or parent, whether friendly and personable, whether he or she gives to charity, and other such characteristics) are largely irrelevant for modern politicians whom almost none of us will actually meet. But the extent to which we rely on personal characteristics that are probably irrelevant for actually performing the job of an elected official is quite noticeable (Masters 1989). Moreover, because we pay attention to such characteristics, politicians themselves emphasize these characteristics in campaigning.

COSTS AND BENEFITS

Many policies provide concentrated benefits to a small number of citizens but impose diffuse costs on many. In many cases, the aggregate benefits to the few are much smaller than the sum of the costs to the many. Farm supports generate large incomes for a few farmers, but all consumers pay higher prices for food as a result. For such a program, the gains to the beneficiaries are smaller than the losses to the losers, because one effect of the policy is to raise prices above costs, and some potential consumers, therefore, reduce their level of consumption of food. This is a cost to consumers—they purchase fewer desired products—but there is no gain to farmers, because they sell less product. This is called a deadweight loss and it is well known to economists. Thus, this program imposes net costs on society. Other examples are easy to find. Tariffs and other import restrictions have the same effect; the gains to domestic producers from higher prices are smaller than the losses to domestic consumers from the price increases caused by reductions in international trade. Unionization and many occupational licensing programs have the same effect. Union members or licensed professionals gain, but the loss to consumers from the increased prices more than offsets these gains. Minimum wages increase earnings for some low-income persons, but others lose their jobs or don't get hired in the first place.

Economists commonly study such programs, with an eye to measuring their costs and also to determine the sources of such inefficiency. The theory of the inefficiency of such programs is a staple of economic analysis, and an analysis can be found in any introductory textbook. The analysis of rent seeking (Buchanan et al. 1980) suggests that the beneficiaries can more easily organize and lobby for benefits, and this is part of the reason for the success of such programs. That is, consumers on average lose such small amounts from each program that it would not pay for them to spend resources lobbying to eliminate the program; indeed, it would generally not pay to spend resources even to learn about the costs. This is part of the explanation. However, it is not the complete answer.

We do not know fully why some interest groups are able to use the political process in this way. Mancur Olson (1965) first raised this issue in a very important work. Olson pointed out that interest groups would be subject to free riding, of the sort that we have seen throughout this book. That is, potential members of an interest group would prefer to avoid the costs of joining and allow others to bear the costs of organizing and seeking benefits. For example, if I am a textile producer, I might like tariffs on imported textiles, but I would like them just as much if someone else bore the lobbying costs of getting them passed, and I would save my share of those costs. Olson argued that as a result groups would be forced to provide selective benefits to potential members—benefits that were privately valuable, such as newsletters or pension funds. Another important implication is that primarily groups of producers (businesses, labor, professions) would organize and that consumer groups would be relatively unorganized. This has been a powerful theory and has revolutionized the study of interest groups by economists and political scientists. (For a recent discussion, see Baumgartner and Leech 1998.)

However, there are still remaining puzzles. For one thing, it is universal that interest groups phrase their arguments in terms of the public good. They argue that their members deserve benefits for some reason—farmers are the backbone of America; steel producers are suffering from "foreign excess steel capacity and market-distorting practices" (from the American Iron and Steel Industry website); consumers need protection from paralegals who might offer to write (allegedly defective) wills at low prices if the American Bar Association did not stop them. In no case does an interest group argue simply that they are big enough to extort money from the political system, and such groups always feel obligated to provide some public interest rationale for receiving benefits. Why is this rhetoric needed?

Moreover, Olson's theory does not seem an adequate explanation for all interest groups. Baumgartner and Leech (1998) point out that people join interest groups because of solidarity incentives, which they define to include benefits that arise from the act of associating with others. They also indicate that members obtain expressive benefits, which are defined as the gains in utility from participating in an interest group. These benefits are not part of the economic analysis of interest groups.

I believe that at least part of the answer is in terms of the power of identified individuals in the political system. When a tariff is being debated, particular workers will expect to lose, and they know who they are. Moreover, others observe them and see that they will lose. Rhetoric often stresses the benefits to these individuals. Potential gainers from reduced tariffs and lower prices are amorphous and anonymous, and so have less weight in political decision making. Similarly, when a union organizes, we can observe wages of union members increasing. These beneficiaries are identifiable individuals. Those persons who are denied jobs because of the higher wages are again anonymous and cannot be seen. Indeed, they themselves probably do not know who they are. Again, the identifiable individuals have a privileged position in the political process.

SAFETY REGULATION

I learned firsthand about the political attention paid to identified individuals when I was the chief economist at the Consumer Product Safety Commission. The agency had a moderately sophisticated statistical system in place for identifying risky products and would systematically look for dangers to fix. This was clearly a Rationality$_2$ program. But about once per year, some freak and newsworthy accident would occur. When this happened, the agency would suspend its ongoing projects and immediately devote attention to that risk—even if, as often turned out to be true, it was a very low probability event. In my years at the CPSC, some of these headline hazards, as we called them, were child entrapment in reclining chairs; death from a lawn dart, in a game that has now been banned; and the danger of exploding cigarette lighters. In all these cases, it was news attention paid to one accident and one identified victim that caused the agency to devote resources to the hazard. In all of those cases mentioned above, the risks were too small to be worth fixing, given that the cost was fewer resources to devote to more substantial risks. Moreover, safer products are more expensive, and so some consumers are priced out of the market by increased safety regulation, but these consumers and their losses are unidentifiable.

The role of identifiable individuals in decision making by the Food and Drug Administration is well known. If a drug is approved and saves lives, those whose life is saved may never know that it was the drug, particularly if the benefit is probabilistic. Many of us walking around today are alive because medications have reduced our blood pressure or cholesterol levels, but we cannot identify ourselves with any degree of certainty. If someone dies because a drug is not available, again no one will generally know that the life could have been saved if a drug had been approved. On the other hand, anyone dying or suffering because of side effects of a drug—an identifiable individual—will know (or his relatives will know) that the drug caused the harm, and such events get large amounts of publicity. As a result, the FDA is overly conservative in approving new drugs. That is, more lives could be saved if the agency approved drugs more quickly with less testing. But some drugs approved would lead to deaths of identifiable individuals, even though on net more lives would be saved, and, therefore, the agency is overly conservative (Peltzman 1973.) This costly decision making is due to our evolved preference of benefiting known individuals at the expense of unknown or statistical individuals.

There is also evidence that we are concerned with deaths of those we know or visualize, but become psychologically numbed to deaths of large numbers of persons (Slovic 2000). This may again be because in the evolutionary environment the notion of one thousand deaths—let alone one million—was not something that could exist, or that our ancestors could grasp. Our ancestors simply would not have known or known about that many people. This may make it difficult for humans to grasp the significance of certain tragedies or of certain political practices, including genocide. After the September 11, 2001, World Trade Center bombing, the *New York Times* ran a series of pictures and brief biographies of the victims. This will help readers in grasping the significance of the event, since the victims are made identifiable individuals by this action.

ADDITIONAL EXAMPLES

Are some consumers deceived in the marketplace? Pass a law outlawing deception. Is unemployment too high? Pass a law putting a quota on imports to preserve American jobs. Do plants sometimes move out of cities without giving adequate notice to workers? Require plant closing notice. Do managers sometimes lose their jobs as a result of takeovers? Make hostile takeovers more difficult. The list is endless. All of these laws (and many others) are aimed at protecting identifiable individuals. Those harmed by deceptive ads, those hurt by dangerous products, unemployed persons, workers who do not receive adequate notice, managers who lose as a result of corporate restructuring—all are immediately identifiable, and the policies appear to directly benefit them. But of course these policies ignore second level systematic effects, with costs borne by unidentifiable persons. Laws regulating deception often have the effect of denying useful information to those who were not deceived. Import quotas and tariffs reduce incomes of consumers. Plant closing notices lead to fewer plant openings, and, therefore, to some workers (unidentifiable) being denied jobs. Laws protecting managers lead to losses for shareholders and to higher prices for consumers but not in any identifiable way.

NEWS STORIES

There is an additional bias in decision making toward known individuals. In thinking about the importance of identified individuals in decision making, remember that our decision-making aptitudes evolved in small societies—probably no more than one hundred persons. Then think about the implications if something happened to a person in our band. Before mass communications, a person probably would learn of a hazard only if it harmed someone in the community. Such risks were likely to be sufficiently probable to be worth worrying about. If there were one hundred people and something harmed or killed one of these people, then the risk of that event was fairly large; the probability of it harming someone was about 1 percent. Thus, it would have been rational for our ancestors to pay careful attention to things that happened to individuals they knew. If a tiger killed someone at a certain location, that would be a location to avoid. If someone ate something that caused them to become sick or to die, that would clearly not be something that we would want to eat. Dunbar (1996) has stressed the importance of gossip in human evolution and in current behavior. Gossip is *one* way of learning about what has happened to identifiable individuals before newspapers, TV, and the Internet, perhaps the only way.

We now live in much larger societies—in the United States, a society of about 275,000,000. If something happens to one individual in the United States, this does not mean that it is something with a 1 percent chance of happening to us. Indeed, a one in a one million chance of something happening is considered quite low—but we would expect 275 of these events to happen each year. If the media bring these events to our attention we may overestimate the risk because we are evolved to pay attention to things that we learn about that happen to individuals. Indeed, the more unusual the death, the more newsworthy it becomes, and we may be more likely to learn about such risks than about more common and more significant risks, such as death from automobile accidents. Even though flying in a commercial airplane is about one hundred times safer than traveling in a car, people are often more afraid of flying, in part because of the publicity given to plane crashes. Because of the publicity given trivial risks, people incorrectly perceive that the world is becoming riskier. In reality, it is becoming ever more safe, and life expectancies are continuously rising. Life expectancy at birth in the United States was 59.7 years in 1930; by 1997 it had risen to 76 years. Moreover, death rates from accidents of all sorts are falling in the United States. Psychologists who study decision making have given a name to this phenomenon. They call it the availability heuristic. The argument is that we pay too much attention to facts that we think of easily—that are available to us in decision making (Kahneman et al. 1982). But the availability heuristic stood us in good stead in the EEA, and it is not surprising that we may sometimes overuse it today.

We may be selected to pay more attention to bad news than to good news. This would explain why the press seems to emphasize bad news (J. Simon 1999). Bad news sells better because it is what we have evolved to favor. If gossip and learning about our neighbors was an important input into our evolution (Dunbar 1996), then one important thing to learn would have been about dangers. By paying attention to things that happened to individuals, our ancestors could have learned to avoid those situations. We also seem to pay particular attention to news stories about bad things that happen to identified individuals. Glassner (1999) discusses the American overreaction to many risks. Most of them are statistically low probability events, but also newsworthy events. Indeed, they are often newsworthy just because they are unusual.

One major example is crime. There are two reasons why we might be selected to pay attention to crime. First, crime is dangerous. If we can learn about how crimes happen we may be in a better position to avoid being victimized. Second, a criminal is a social cheater; that is, a criminal is someone who is playing defect in the great social prisoner's dilemma. We have been selected to pay particular attention to social cheaters to avoid being victimized by them. Thus, it is not surprising that news stories about crime, and even fictional accounts of crime, are important for humans. (I have always been saddened by observing that there are more television shows about police than about economics professors.)

Another example is associated with the dreadful shootings at Columbine High School in Littleton, Colorado, in 1999. These shootings were tragic, and horrible. But at the same time, the number of high school students expelled in academic year 1997–1998 for bringing a firearm to school fell by almost one-third (Kronholz 1999). Moreover, youth homicide and violent crime rates have also been falling (Winslow 1999). In spite of the statistical evidence of actual reductions in school crime, the headline event of Columbine High School led to many changes in policy. In Houston, for example, the school district began random metal detector tests; Montgomery County, Maryland, now requires identification badges for all students and staff (Kronholz 1999). This episode is an example of policies being driven by news headlines involving identifiable individuals rather than by more meaningful but more difficult to interpret statistical evidence. In an ongoing series, the *Wall Street Journal* online has documented many harmful effects from school zero tolerance policies caused by these shootings.

What implications does this analysis have? We pay attention to the wrong kinds of events. Those of us concerned with correct decision making should train ourselves to pay more attention to underlying facts and less to particular events. In science, anecdotal evidence is treated with great care. But in policy making it is often quite important.

Self-Deception

I argued in chapter 3 that people often want to fool others and that one way to do so is to first fool oneself. That is, a person who is convinced that he is telling the truth is a better liar than one who knows that he is lying. As a result, we engage in substantial amounts of self-deception (Trivers 1971).

One implication relevant for political analysis of this self-deception is that humans (acting as individuals or as members of interest groups) wanting special favors from the government can easily convince themselves that these favors are actually in the public interest. They convince themselves that the benefits are not just for the private benefit of the interest group. Farmers argue and probably believe that farm supports are good for the country, not just for the farmers. Domestic textile manufacturers and workers argue and probably believe that tariffs on textiles are for the good of the country. Unionized workers argue and probably believe that immigration restrictions are for the good of the country. Barbers and lawyers argue that restrictions on the practice of haircutting or law are in the interest of consumers. Toxicologists working for industry perceive chemicals as less harmful than government or academic experts (Slovic 2000). As an economist, I believe that the National Science Foundation is good for the country, and I can present theoretical reasons for believing this. Nonetheless, I tell my students that the arguments might well be biased because the NSF is the scientists' subsidy, and I would have incentives to deceive myself about the benefits it provides.

The implications are that, if we are seeking truth in a scientific sense, we should be very careful about believing anything that is in our own interest. Just because we may really be convinced that something is true does not mean that it is true—and especially if it is in our interest to believe it.

Status Quo Bias

One finding of the cognitive psychologists that will probably turn out to be correct is what is called status quo bias. This is a cluster of findings, each sometimes given their own name. We exhibit an endowment effect—things we actually possess are valued more highly than makes sense. For example, the amount we demand to sell a possession is greater than the amount we are willing to pay to buy the same good. We also exhibit loss aversion—that is, losses are overvalued relative to gains. (Discussions of these issues are widely available; see for example Rabin 1998.)

Endowment Effects

Rational theory indicates that willingness to pay and willingness to sell should be the same. That is, if a consumer would pay at most $10 for an item (and thus values it at no more than $10), then he should be willing to sell that item for $10. This tenet of rationality seems to be widely violated in experimental settings. This violation is called the endowment effect. One well-known example involved giving mugs to randomly selected students at Cornell. Students given a mug valued it at about $7.00; other students would pay no more than about $3.50 for the same mug (Kahneman et al. 1990). This has been widely discussed in the literature (e.g., Curran and Rubin 1995).

Mass produced standardized items are very new in human experience. During the EEA and even more recently, produced items would have been idiosyncratic. A well-defined market for a standardized item would not have existed: Walmart was not part of the EEA. Therefore, it is plausible that people would prefer items they already had to the opportunity of acquiring a new item that might not have the same characteristics. Indeed, for many goods even today markets are thin. Moreover, to the extent that items are idiosyncratic, then search costs and other transactions costs of obtaining desired items may generally be significant.[4] In the experimental situations where the endowment effect seems to operate, subjects may simply be transferring a generally efficient rule to a context where it is not efficient. The decision costs of tailoring the rule to each specific situation may overall be greater than the costs of the occasional error from misapplying a rule, as discussed in an interesting paper including considerations from both economics and evolutionary psychology: Samuelson 2001. This may be particularly likely in artificial situations designed to create errors.

If this is so, then in pure financial markets or markets where goods have no consumption or idiosyncratic value, endowment effects should not be observed. This seems to be the case. For example, Roth (1995, 97, note 111) indicates that buying-selling disparities are smaller when individuals are acting as agents for others. Thaler (1992, 72, note 3) indicates that loss aversion does not apply to commercial transactions: "Loss aversion is expected to primarily affect owners of goods that had been bought for use rather than for eventual resale." The theory advanced here indicates that endowment effects should be smaller when abstract goods or financial goods are being traded then when goods with innate consumption value (such as cups) are being traded. Moreover, this also implies that endowment effects should not have significant impacts in market settings, where there are well-defined prices.

Several have proposed an alternative theory (see, e.g., Ellingson 1997 and Huck et al. 1999). The argument is that we have evolved preferences that help us bargain. The EEA was a world with few organized markets with well-defined prices and so most exchanges would have been negotiated bargains. Then individuals exhibiting an endowment effect would have achieved better outcomes for themselves than those who did not overvalue goods that they possessed. While this theory is consistent with the evidence and can be shown to work, given the assumptions, it depends on substantial amounts of bargaining and exchange occurring in the EEA—which, as indicated in chapter 1, may not have been the case.

In any event it appears that we do exhibit an endowment effect. This has several implications for political preferences. Because many of these effects are related to loss aversion, I will discuss them together below.

Loss Aversion

Loss aversion means that we dislike losses much more than we like gains— about twice as much. A loss of $100 creates twice as much disutility as the gain in utility from a gain of $100. This is different from normal risk aversion. Loss aversion can be explained if there is a kink or bend in the utility function at the current level. To the extent that this is true, it is inconsistent with economists' normal models of utility maximization. For example, Rabin (1998) indicates that most people would refuse a fifty-fifty bet of losing $10 or gaining $11. If so, then people are forgoing opportunities to increase wealth, for accepting such bets would be a way to increase wealth. Moreover, people seem to weigh utility in terms of changes from the status quo, rather than in terms of actual levels of utility.

If we recall that evolution would have selected individuals to maximize fitness, these behaviors become understandable. Assume that in the EEA people would generally have been at subsistence. More specifically, people would have had as many children as they could support, so they would have essentially put themselves at subsistence through Malthusian behavior. Then any reduction in wealth or income (in a world with no storage, they may be the same) would have led to starvation of a child. Any increase in wealth might have led to an additional child after some time had passed, and the fate of the unborn child would have even so been doubtful. Thus, individuals who accepted fair gambles in these circumstances, or who did not overweigh losses, would have been less likely to leave offspring, so there would have been selection pressure for exactly the overweighing of the status quo that we observe. This is consistent with risk-sensitive foraging theory, which argues that there is a minimum need for calories and that below that minimum death (and a consequent loss of fitness) occurs (see Rode et al. 1999; Rode and Wang 2000).

This analysis suggests that degrees of status quo bias should vary with demographic characteristics. For example, young males with no children might be more likely to be risk seeking and older males and females risk averting (Rubin and Paul 1979; Robson 2001). These implications are testable.

The evolutionary arguments do not lead us to reject status quo bias. Rather, they suggest that this bias might be part of actual utility functions. However, if so, then this is not a sign of error or irrationality. Just as earlier students of human behavior were forced to replace expected value with expected utility as a result of experiments with the St. Petersburg Paradox, so it may be necessary to replace maximization of expected utility with a paradigm where the status quo is privileged.[5]

Implications

Loss aversion and status quo bias have several implications for political preferences. One is that it will be difficult to change an existing government program. Any program creates beneficiaries, and these beneficiaries will strongly resist any change that reduces their incomes. This resistance will be stronger than might be expected based on the amount of potential loss. Moreover, even if the program has potential gains for the losers (e.g., offers to pay compensation), the losers may still resist because the sure losses are overvalued relative to the potential gains. This means that decision makers should be very careful in imposing new programs because the political cost of later eliminating the programs, if it should become worthwhile to do so, will be higher than might appear.

Framing Effects

One finding from the cognitive psychology literature is that the way in which a problem is posed has important implications for the answer. For example if a problem is posed in terms of losses, then the preference for loss aversion kicks in, and we get one answer. If the same problem is posed in terms of gains, our minds might give another answer. (Rode and Wang 2000 show that this inconsistency disappears if a problem is of the sort that would have occurred in the EEA.) Consider credit cards. The contracts credit card providers have with merchants forbid merchants from charging a premium for the use of the credit card. But these contracts allow discounts for cash. Clearly, the two behaviors are identical, but just as clearly, credit card providers believe that consumers will view them as being different. This is an example of a framing effect. This is contrary to economic models of rationality, in which the answer is independent of the form of the question.

Framing effects are simply the result of applying particular modules to decision problems. If different frames elicit different modules, we will observe framing effects. Moreover, if questions are asked in such a way as to trick the mind into sending a problem to the wrong module, we will get the wrong answer.

With respect to political decision making, the theory of modularization and the existence of framing effects means that the way in which a political problem is posed will have important implications for its answer. Politicians know this, and one important part of political debate is over the way in which issues are framed. If one politician can frame the program of another as a loss, then loss aversion kicks in and people may be opposed. The politician advocating the program will of course try to frame the issue as a gain. For example, in the recent (2000) United States presidential election, Governor Bush, in his quest for the presidency, proposed putting part of Social Security contributions into the stock market, which he stressed would provide larger payments (a gain). Vice President Gore characterized the same policy as risky, stressing that there might be an associated loss. It would be useful if citizens could understand these issues and consider the net effect of programs on their welfare, but our natural tendency is to view policies thorough particular frames.

Motives

It is common to judge the desirability or undesirability of an action by the motives of the actor. This method of decision making may have made sense in a simple environment with few persons, such as in the EEA. In a large complex environment such as the one we inhabit today, motives are less important than outcomes. This is particularly important with respect to my field of study, economics. At least since Adam Smith, economists have known that selfish motives can lead to desirable social outcomes—the invisible hand. But this lesson is counterintuitive and not part of our evolved mental architecture. As a result, people too often judge the moral worth of an act by the motives of the actor, rather than by the outcome. This may be part of the reason for the hostility of intellectuals towards a capitalist system and the partiality of many intellectuals toward government solutions to perceived problems. Capitalists base their actions on profits, and government alleges that it is acting in some sort of pubic interest.

An example of this fallacy is presented in Singer (2000, 44). In his call for Cooperation instead of Competition, Singer characterizes contemporary society: "America in the twentieth century has been the paradigm of a competitive society, in which the drive for personal wealth and to get to the top is widely seen as the goal of everything we do." But what Singer misses is that the method of obtaining personal wealth and the way to get to the top is by producing goods and services that others find worth buying. That is, a capitalist society rewards those who produce things that provide benefits to others. This is a subtle and counterintuitive notion, but one might hope that an ethicist with the reputation of Singer, a well-known chaired professor at Princeton, could have grasped it. The fact that he could not explains the difficulty others have with this concept. In this same passage, Singer (2000, 45) also commits the zero-sum fallacy discussed in chapter 1: "I doubt that the greater wealth for the middle and upper classes could make up for the increase in human misery that this has brought to the poor." But he provides no evidence—and could not provide any evidence, since it is not true—that the increase in the wealth for the middle and upper classes has in any sense come at the expense of the poor.

This confusion of motives with results may be in part due to the sort of zero-sum thinking discussed in chapter 1. In a zero-sum world, more for me means less for you, and the terms of a transaction are very important. In a positive sum world, we can both gain from exchange, and motives become less important because we can both gain. But zero-sum thinking would lead to a devaluation of the benefits of exchange, and so to a criticism of capitalism. It is ironic that many intellectuals, who pride themselves on seeing below the surface and understanding the deep structure of society, are in fact confused and focused on the surface appearance of trade and exchange, rather than on the true benefits.

Juries

Most of the analysis has been of electoral politics. However, the analysis of this chapter has some relevance for explaining the behavior of juries. Individuals serving on juries are in virtually identical situations as experimental subjects. Moreover, jurors are in a situation that would almost maximize the chance of reaching erroneous or irrational decisions. Psychologists first provided experimental evidence of errors in decision making. Economists were skeptical of these results and attempted to replicate the experiments to correct for what they perceived as misspecifications. Camerer (1995) indicates that the main differences are that psychologists use natural stimuli, do not pay subjects, and do not repeat tasks. Economists pay subjects, prefer blandly labeled random devices as stimuli, and insist on repeating tasks.

Now consider a jury. It makes a decision about an accident that has already (really) occurred, the essence of a natural stimulus, and moreover often one with a substantial emotional load. Jurors are not paid for correct decisions. Finally, cases and juries are unique, so there is no chance of a juror repeating the situation and learning of correct answers. The efforts of experimental economists were aimed at finding treatments that would lead subjects to make more rational decisions. The changes in treatment used by economists went some distance (though not all the way) toward achieving this goal. Thus, to the extent that the institutions governing juries are like those psychologists use and unlike those economists use in experiments, then these institutions are more likely to lead to relatively less rational decisions and to more cognitive errors.

Juries seldom have a chance to consider explicit cost-benefit analysis. There was one famous case where a jury was confronted with exactly such evidence, and the result was a public policy disaster. This was the Ford Pinto case.[6] Schwartz (1991) persuasively shows that juries are totally unwilling to accept any hint of an analysis explicitly measuring the cost of lives saved. This is true even when the law expressly requires such balancing. It was the fact that Ford had undertaken such calculations that induced the jury to award large punitive damages in this case. Moreover, the strength of the jury's ire was sufficiently strong so that virtually no defense lawyer is willing to make arguments relating to cost. Schwartz's distillation of conversations with several defense lawyers contains the following statement:

> However, one argument that you should almost never make is that the manufacturer deliberately included a dangerous feature in the product's design because of the high monetary cost that the manufacturer would have incurred in choosing another design. If you do argue this, you're almost certain to lose on liability, and you can expose yourself to punitive damages as well.

But arguments regarding the cost of safety improvements are the essence of efficient, rational decision making with respect to safety.[7] If such arguments cannot be made before a jury, it is unlikely that juries can efficiently act as agents for consumers and unlikely that reliance on juries will lead to correct outcomes. Indeed, because of the results of *Ford* and the ability of plaintiffs to obtain documents through discovery, this case has probably meant that many firms are unwilling to even undertake such cost benefit analysis for internal planning purposes, let alone to make such arguments before a jury. In this sense, *Ford* has probably lead to excess harm and injury in the economy. The unwillingness of juries to consider such evidence may be a framing effect, the result of unwillingness to shift from a safety frame to a money frame. Policy makers and others using Rationality$_2$ reasoning make such decisions routinely, but others resist them.

Many scholars, in work summarized and extended in Hanson and Kysar (1999), have argued that cognitive effects would lead to biases in decision making so that consumers would likely make errors in signing contracts.

They used this argument to justify interference with contractual freedom and court intervention in private transactions. But interference with free contract will imply that more contractual disputes will go to juries. Juries are more likely to be subject to bias than are consumers. Consumers sign many contracts and buy many products. Consumers pay directly whatever costs are associated with errors, and so receive feedback from erroneous decisions. Thus, consumers have a chance to learn from any mistakes they may make. Jurors have no such chance. Jurors (as consumers) may end up paying higher price for goods and services because of their decisions as jurors, but the link is neither obvious nor immediate; it is presumably a link that can best be understood using Rationality$_2$ type reasoning. As a teacher of law and economics, I can say that it is not intuitively obvious to law students either. Thus, if we believe that cognitive biases are important in decision making, we must believe that they are more severe with respect to jurors than with respect to individuals engaging in exchange for their own benefit.

Expected Jury Biases

What particular biases might we expect from jurors in product liability matters? In general, the predictions all point in the same direction: juries are likely to award damages more often and award higher damage payments than consumers would desire ex ante. This additional level of payments will arise partially through a greater likelihood of finding liability and partially through awarding larger damages than would be desired ex ante. Since the errors discussed are systematic, there is no presumption that awards would be random, and a finding of predictability in pain and suffering awards would not be inconsistent with the theory proposed here. Awards will be predictable and they may be internally consistent, but they will be biased upward relative to efficient levels.

Consider first liability. In a negligence system (and in the odd version of strict liability that governs the United States current product liability system) a firm is negligent if it does not take all cost-justified precautions. But whether a precaution is cost justified depends in part on the probability of the harmful event occurring. A jury observes the product after an accident has happened and must then attempt to infer what level of precautions would have been efficient when the product was made and sold. There are reasons to expect that juries will form incorrect estimates of the relevant probability.

One common cognitive error is hindsight bias, discussed earlier. Once an event has occurred, then subjects and presumably jurors as well view the probability of that event as being greater than before the event occurred. Thus, a jury, faced with an already existing accident, will believe that the probability of the accident ex ante was greater than may be objectively true.

Therefore, even if a firm behaved nonnegligently and took all cost-justified precautions, a jury may find negligence because of its overestimate of the probability of the accident. There is both actual (Arkes and Schipani 1994) and experimental (Kamin and Rachlinski 1995) evidence of the importance of hindsight bias in a litigation context, leading to excessive liability. Kamin and Rachlinski show that experimental subjects are likely to find liability after an event has occurred in a situation where subjects viewing the same situation before the event find precautions not worthwhile. In their experimental treatment, only 24 percent of the ex ante subjects found taking a precaution worthwhile, but 56 percent in the ex post situation found negligence for failure to take the same precaution.

A similar result follows from what has sometimes been called the law of small numbers (Camerer 1995). This is the tendency of experimental subjects to overgeneralize from small samples, a form of Rationality$_1$ decision making. Thus, if the major relationship jurors have with some product is observation of the effects of a mishap regarding this product, the jurors might well view the product as being more dangerous than it is, again giving rise to overestimates of the risk associated with the product.

As for damages, it is very likely that juries will find greater levels of damage payments when viewing an accident ex post than consumers would have wanted to contract for ex ante (see Rubin 1993). We have seen that losses are overweighed relative to gains. Before an accident (ex ante), both the accident and the payment for the accident may be viewed as losses (If you are injured, you will lose $10,000 in wages and medical payments. If you are covered for these injuries, you will pay $10 more for this product.) Thus, it may be that consumers will weight gains and losses approximately correctly. Moreover, consumers ex ante are accustomed to making exactly these calculations routinely in markets. But after the accident (ex post) the actual suffered loss will have excess weight. Losses, not product benefits or money saving from reduced prices, are the entire focus of the trial. Indeed, the evidence from *Ford*, cited above, suggests that jurors are unwilling to consider the ex ante perspective (i.e., the perspective of the cost of additional precautions). Therefore, cognitive theory indicates that juries will systematically place more weight on the actual accident and award more in damages than consumers would desire ex ante.

Some experimental evidence demonstrates this. Calfee and Winston (1993) have shown that ex ante consumers are not willing to pay much for compensation for pain and suffering, as the theory would predict. On the other hand, experimental studies of ex post compensation for pain and suffering report, consistent with cognitive theory, that the frame in which the problem is set determines the outcome. McCaffery et al. (1995) test an additional implication of the endowment effect, the difference between willingness to pay and willingness to accept. Strong experimental evidence shows that consumers will demand more to give something that they own up than they would be willing to pay to buy the same item. In an injury context, this implies that if jurors are asked to award compensation based on the ex ante (selling price) perspective, values should be larger than if the award is framed in terms of the ex post (making whole) perspective; and this is what they find. Of course, both amounts provide more insurance and compensation than the true ex ante amount that theory would predict consumers would be willing to pay for, and more than is consistent with the experimental evidence provided by Calfee and Winston.

The identifiable victim effect will also work to increase damage payments from a jury. After an accident, the victim is identifiable; indeed, he (or his heirs) is in court. Before the accident, the chance of harm is to a statistical or unidentified individual. But we saw above that people tend to overweight harms to identifiable individuals relative to harms to statistical individuals. This will also serve to increase damage payments.

Summary

Men and women differ in risk preferences, and these differences have implications for political behavior. As a result of greater aversion to risk, women seem to prefer larger government. Because war had benefits for some males but mainly costs for females, women prefer reduced levels of combat.

In the rest of this chapter, I discuss decision-making mechanisms. As we have gone from groups of perhaps fifty individuals to our huge societies of today, we have adapted decision-making mechanisms from the smaller societies. This adaptation has worked well, at least in western societies, but not perfectly. We can see traces of earlier ways of making decisions. However, I also show that some authorities (many cognitive psychologists, and more recently, some economists as well) may have overestimated the extent to which humans make cognitive errors. For an important class of decision problems, individuals do poorly if questions are asked about probabilities, but much better if the same questions are phrased in terms of frequencies. This makes evolutionary sense, as it is possible to observe frequencies and make decisions on this basis.

A useful way to characterize human decision making is in terms of two separate mechanisms. One mechanism is unconscious and makes use of all available information in an environment; the other is more formal and satisfies self-conscious verbalized rules. These mechanisms may explain some differences between opinions of experts and others on various policy issues. Humans are quite good at making decisions in many contexts using available information, and fast and simple heuristics can reach surprisingly good results. Decisions also exhibit various forms of status quo bias, including loss aversion and endowment effects that probably had value in the EEA but may be less valuable now. For example, loss aversion might make it excessively difficult to terminate a program even when it should be ended. In many political decision problems, we pay excessive attention to identifiable individuals. In the EEA, there were only identifiable individuals, and we did not evolve to be able to make good decisions for large groups of anonymous individuals. This can lead to some errors in policy.

Framing effects, which may be a result of use of the wrong decision module, can also lead to errors. People can deceive themselves; in politics, this self-deception probably takes the form of individuals convincing themselves that politics that benefit them are socially desirable. I also consider jury decision making. Jurors are likely to be subject to many cognitive errors in a situation where there is little cost or benefit to making correct decisions. In the context of liability suits, this probably means that juries find liability too often and award excessive damage payments to victims.

CHAPTER 8

Relevance of the Pleistocene for Today

We evolved in groups of about fifty individuals, and our political preferences evolved with us. We now live in political groups of some tens or hundreds of millions. We have the same evolved political preferences that
developed in small groups of individuals. It is truly amazing that we have
been able to adapt our evolved political behaviors to get them to work in
such large agglomerations. I first discuss some analytic implications of the
analysis and then some policy implications.

Analytic Implications

Philosophers have spent a good deal of time and ink debating the relationship between what is and what should be. I do not want to enter into that
debate. I assume that our preferences have evolved, and the issue I address
here is the way in which we can structure institutions so as to satisfy those
preferences more effectively. (For a similar argument by a philosopher, see
Arnhart 1998.) My approach is basically utilitarian, as is consistent with the
argument in chapter 4. However, I advance the argument because I argue
that the utility functions that go into the utilitarian calculus are those that
have evolved.

The first thing to note is that all political systems are not created equal.
Humans do have a particular and specific set of preferences, and some systems are simply better at satisfying those preferences than are others.
Edgerton (1992) and Kronk (1999) describe tribal societies with institutions
that are harmful—sometimes extremely harmful, and even fatal—to their
citizens. For more advanced societies with institutions that benefit powerful
elites and dominants, and harm many others, we need look no farther than
the Nazi or communist regimes of the last century, which each caused tens
of millions of deaths (Rummel 1994; Glover 2000). I do not take the stance
of cultural relativism that is common among anthropologists and others (as
described, e.g., in Kronk 1999). I believe that humans have sufficiently common preferences so it is possible to say explicitly that one society is better
than another—not based on my individual preferences or prejudices but
rather based on general human preferences. Harmful cultures may be those
that are organized to benefit dominant individuals at the expense of others,
as has often been true of human societies.

There are substantial difficulties in comparing societies with respect to human happiness. Indeed, Kenneth Arrow (1970) won the Nobel Prize in Economics in part for demonstrating that under certain circumstances such comparisons are impossible because there is no meaningful way of aggregating human preferences. However, I am here dealing with aggregation at a much broader level. For example, one assumption for Arrow's proof is that dictatorship does not exist in society. But one difference between the societies I consider is exactly the issue of dictatorship—I consider that nondictatorships are preferable to dictatorships. Other comparisons are at this level, and the rankings I provide would bring about little debate, if one is willing to accept the notion of such rankings.

I believe that the current American and western political systems have many desirable properties. And, I believe that they are the best systems that have existed, in the sense of being the most consistent with evolved human preferences and the best at satisfying these preferences. But even these systems have in the recent past had substantial problems. It was only 150 years ago that the United States had slavery, and it was only fifty years ago that many parts of the United States engaged in exceedingly unpleasant racial and ethnic policies. Although such policies are no longer legal here, some individuals still behave in racist ways. It has only been since the passage of the Nineteenth Amendment in 1920 that women were granted the right to vote in the United States, and the Civil Rights act enfranchising racial minorities is even more recent. Further improvements are of course possible, and I do not claim that any system is perfect. Indeed, much of my professional writing as an economist is aimed at addressing what I believe are flaws in the United States political and legal system.

I have identified several dimensions of political preferences and behavior. One important point of this book is that in almost all dimensions, modern western democracies, and particularly the United States, do a better job of satisfying these preferences than any other human society of the past or present. I discuss this with respect to each dimension addressed.

Individualism

Humans are individualistic. Because of frequency dependent selection and perhaps for other reasons, we are different from each other and have different preferences. Evolution will not generate the same sets of genes or preferences in everyone. Human individuality is important, and political ideologies or theories that assume everyone to be the same will invariably create much human misery. Humans have different genetic endowments and choose different strategies, which is an argument for different tastes. These differences also explain why individuals want some freedom from social control.

We live in large anonymous societies. In these societies, there are numerous possibilities for consumption, and, therefore, we can satisfy these different tastes. Similarly, in modern capitalist societies the division of labor is extremely fine. As a result, there are many jobs and occupations individuals can fill. This again gives us possibilities to satisfy individual preferences. In earlier societies, consumption possibilities and occupational differentiation were limited, so humans had fewer opportunities to satisfy unusual or unique preferences. There was often less privacy, and individuals were forced by social pressure to conform (Posner 1980). Some societies have attempted to enforce conformity and suppress differences. This is not a characteristic of modern western societies. For example, in the United States and elsewhere, homosexuality was until recently condemned. Laws are no longer enforced against this behavior, and society is increasingly protecting homosexuals from various forms of discrimination. More generally, the contemporary United States allows widely varying life styles. The mobility that capitalism created also allows individuals to move within society to find a congenial life style, and geographic mobility within a given political jurisdiction is relatively easy and is apparently becoming easier.

Sociality

Humans are a social species. As humans, we have always lived in groups. In the evolutionary environment, these groups would have been mainly genetic kin, but this is not the only mechanism for group identity. Group identity today is much more flexible. Humans seem to be able to easily identify with many groups, with many definitions, including even completely arbitrary identification mechanisms. This flexibility of the group identity mechanism must be counted as a desirable characteristic of humans. In today's society, it is possible to join any number of groups, with interests tailored to virtually any set of preferences. There are groups meeting almost any human desire, and 80 percent of individuals belong to at least one group; many belong to several. Modern transportation and communication, products of capitalist production technologies, make finding and meeting with like-minded others much easier; the Internet, an invention of modern society, is particularly useful in finding and communicating with others. There are few restrictions on forming or meeting in groups; indeed, in the United States the First Amendment to the Constitution protects the "right of the people peaceably to assemble."

Conflict

The negative side of our group preferences is the possibility of ethnic polarization. One group, based on ethnic or racial characteristics, can engage in xenophobic predation against other ethnic groups. The United States has the advantage of not having any large majority ethnic group. Being an American is totally a political, not an ethnic, identification, in that many Americans are only very distantly genetically related to other Americans, which makes the risk of predation by one ethnic group against another much less likely here than in other countries. This appears to be an advantage of the United States over even Western Europe, which also includes modern liberal democracies. After all, the Nazis were in power in Germany only sixty years ago. Such predation has been a real risk throughout most of human existence. More generally, democracies are significantly less likely than other forms of government to engage in ethnic predatory policies (Rummel 1994). Additionally, as incomes increase, such predatory policies also become less likely (Scully 1997).

Although conflict is a part of modern life, for most citizens, the chance of dying in a war is small. For example, in World War II, only 17 percent of the United States male population was mobilized, and only about 40 percent of those served in combat units. Moreover, wars are relatively rare for major societies such as the United States—apparently on the order of about once per generation. In the twentieth century, less than 1 percent of the male population, and even less of the female population, in the United States and Europe died as a result of war (all data from Keeley 1996). I do not want to be overly optimistic—at the beginning of the twentieth century, it was commonly thought that the era of war was over and that humanity was entering an era of peace (Glover 2000), and of course what might be a major war against terrorists has begun in 2001. However, even so, for the average citizen of a western country, the risk of death in war is now quite low and clearly much lower than in the EEA. Keeley estimates that the rate of death in tribal societies from war was twenty times larger than in the twentieth century.

In the United States, we now allow immigration of people from many different ethnic backgrounds. We would gain by allowing even more immigration, but it is important to note that, unlike in the recent past, migration is not racially limited or controlled. Race-based policies such as affirmative action increase the risk of ethnic conflict, however.

Altruism

Humans are altruistic, at least to some point. We are willing to transfer resources to those with fewer resources. This altruism is not unconstrained: we are more willing to transfer resources to those who are temporarily lacking and to those who are ill. We are less willing to transfer resources to those who cannot be expected to become productive or to those who are viewed as shirkers. But our altruism is a characteristic that most would view as a nice property of human nature. In the West, resources are devoted to increasing the consumption opportunities of the poor. Many believe that insufficient resources are used in this way, but the issue of the magnitude of transfers is less important than their existence and form. However, it is not useful to measure success in redistribution by looking at equality. The important issue is the level of consumption of the poorest, and by world standards, the poor living in rich countries such as the United States are quite well off. Recent changes in United States welfare policy, attempting to induce recipients to work, are consistent with evolved preferences for redistribution and for limiting the amount of redistribution to free riders. These changes have led to increasing amounts of resources being transferred to the working poor, as voters no longer perceive that recipients are free riders.

Cooperation

We are also a highly cooperative species. Cooperation in modern society works at two levels. We directly cooperate and work with others in relatively small groups where everyone knows everyone else. This is the evolutionarily old form of cooperation, and the one we intuitively understand best. But the impersonal market is a form of cooperation as well and is a relatively new form of cooperation. It is the form of cooperation studied by the discipline of economics. But it is the most powerful mechanism available for satisfying human wants and for inducing cooperation. In many respects, almost the entire world is engaged in cooperative endeavors, coordinated through the market. Cooperation through the impersonal market is relatively new in evolutionary terms. Therefore, many do not intuitively understand it and are excessively hostile to the market. This hostility may be based on our evolved predilection for envy, or for judging actions based on motives rather than on outcomes. But the modern West gives great scope to relatively unregulated markets, and the effects are greatly increased wealth to satisfy human desires.[1]

Political Power

Humans, and particularly human males, seek political power. But humans do not like to be dominated by others. This tension has been an important part of the human story. In hunter-gatherers, there is a reverse dominance hierarchy, in which subordinates cooperate to keep would-be dominant individuals from becoming too dominant or powerful. As humans moved from foraging to sedentary societies, the power of dominants and elites greatly increased, and this was the situation through most of recorded history. Indeed, recorded history is mainly the story of conflicts between elites, with an occasional rebellion by subordinates (who would generally become dominants if they won) thrown in.

It is in the area of political power that our society performs best relative to other societies. Government is weaker in advanced western democracies than in other societies. The only competitor may be the band level societies of the EEA, which had reverse dominance hierarchies that limited the power of dominant individuals. But in these societies, men generally had much more power than did women. That is no longer true in western societies. More recently, and particularly in the West, we have returned to a world with relatively little power given to dominants. Many mechanisms limit the power of dominant individuals. Governments are designed and planned explicitly to limit such power, and this is the stated purpose of the U.S. Constitution. Moreover, there are many competing power centers in advanced societies, such as powerful businesses. Individuals can compete in any number of hierarchies, not merely in a governmental hierarchy. Since women now have full political equality in most of the West, it is true that today humans in the developed West have more freedom than has ever been true of the mass of humans in any society in the past. Moreover, the recent defeat of the communist countries in the cold war indicates that societies with more freedom are likely to prevail in today's world.

It is a puzzle, however, to explain why many, including many intellectuals, want to increase the power of the government; reducing and limiting this power is one of the great human achievements. Nonetheless, although we may debate the scope of government on the margins, overall democratic western governments give less power to dominant individuals than any other form of government.

Economics

Modern western society is the richest society that has ever existed. Even relatively poor people in the West have access to goods and services that would have been unimaginable even to the rulers in the relatively recent past. Such goods include, for example, entertainment on television, easy and inexpensive travel, and a wide variety of high quality foods available all year long. This level of wealth is itself a real benefit to humans. Frank (2000) has recently argued that much wealth is spent on status seeking and, therefore, does not increase human happiness. Even if true, however, this argument applies mainly to marginal activities in rich countries; Frank does not deny that much of the increase in wealth has gone to buy improved nutrition and increased life expectancies, and is therefore beneficial.[2]

Our level of wealth and technology also makes many political activities possible. One example is the ease of communication and the consequent difficulty of government censorship. Books and videocassettes are one example; the Internet and electronic communication are another. It is also cheaper and easier to utilize the ultimate antigovernment mechanism, emigration, because transportation costs are much lower than in the past.

Nevertheless the unwillingness of many countries to receive immigrants limits possibilities.

The nature of wealth in modern economies is sometimes misunderstood. As discussed in chapter 4, most wealth in modern societies is created through productive activities. Market hierarchies are different than political hierarchies in that markets derive their power from the willingness of consumers to purchase products and services, and firms have no ability to coerce consumers, unlike government hierarchies. Moreover, there is a good deal of economic mobility in modern societies—more even in the United States than in Western Europe (Alesina et al. 2001).

Religion

Humans are a religious species, and religion is an important input into politics. The origin of religion is probably in anthropomorphism—we attribute intelligent agency to many actions where such attribution is inappropriate. Since religious belief is nontestable in a scientific sense, there is no objective method of choosing which religion to believe in. However, some religions are more successful than others in that their adherents are more successful. Christianity and Islam have in common that both are proselytizing religions. That is, both divorce membership from ethnicity, which is an important explanation for their success. Moreover, both impose certain behavioral standards on their members. One duty is honest dealings and honoring of promises and contracts between members, which facilitates exchange and increases the wealth of members. Another is that both regulate certain harmful behaviors, such as consumption of drugs. These characteristics probably explain the success of these religions.

Christianity has long outlawed polygyny, but this is legal in the Moslem religion.[3] Polygyny creates a class of unhappy males, and a powerful government is needed to suppress these males. This may explain the relative lack of freedom in Islamic countries. Because powerful males benefit from this system, it is difficult to change. In the United States today, abortion is freely allowed, although this issue is controversial. By giving women an additional element of control that has been lacking in many societies, we must count this as a net benefit for freedom in our society.

In the United States there is no state support of any religion, which adds another dimension to human freedom—we can choose any religion, or none at all. This is an additional advantage of our society.

Policy Implications

One important lesson involves methods of decision making. In particular, we pay too much attention to incidents involving particular identifiable individuals and not enough attention to statistical and other more objective data. This leads to many poor social decisions. Greater reliance on objective data would in many circumstances lead to better decisions. Moreover, much decision making based on identifiable individuals leads to policies increasing the size and power of government, as government is enlisted to benefit indefinable persons at the cost of imposing unnoticed but real harms on the mass of society. When we as individuals advocate particular positions, we should also be aware that we are rather good at self-deception, and if a position benefits us, we should be skeptical of our belief in that position.

We are altruistic, and so transfers to the poor are not against our nature. But we do not want to undertake such transfers indiscriminately. Policies conditioning such transfers on work effort will be politically more desirable and have actually lead to more being transferred with less political disagreement.

We have some tendency to be envious of others. This is understandable in evolutionary terms. In the evolutionary environment, those with wealth could use it to engross additional females. Moreover, acquisition of wealth may have come from social cheating. But in a world of capitalism and monogamy, the way to become wealthy is to be productive and create benefits for all. We do not live in a zero-sum world. There are many possibilities for increasing human wealth. This means that the wealthy become wealthy mainly by producing things than benefit all of us. Moreover, we live in a monogamous society. This means that the wealthy are no threat to the fitness of the less wealthy, because they cannot engross a disproportionate share of women. As a result, envy is now a socially useless feeling. In the United States, the political system does not and should not rely on envy. Redistribution should be for the purpose of benefiting the poor, not to harm the rich. Capital is productive, and taxes on capital will reduce investment and therefore reduce incomes, but we tend to like such taxes (such as the corporate income tax).

Most governments, including the United States, have active policies to regulate private behavior, in areas such as use of drugs. When I began this book, I was convinced that such policies were misguided. However, after studying sexual selection and handicap competition, I have concluded that there may actually be a justification for them.[4] Nonetheless, the issue is worth further study. However, if we view consumption of such goods as a form of handicap competition, some policy implications become obvious. For example, government propaganda aimed at reducing consumption of these products should not stress their harms, for this merely reinforces their use as signals. Rather, this propaganda should stress that only losers will use this form of competition.

The analysis explains the current debate over abortion policy. Those opposed to legalization of abortions are invoking preferences for larger populations that might have been useful in the evolutionary environment but are no longer needed. Principles of personal freedom would then argue that restrictions on this behavior are not justifiable.

One of the most difficult tasks for economists has been to explain to others that free international trade is beneficial. I must say that we have apparently done a good job of this, and barriers to trade are lower now than at any recent time. But an analysis of the biology indicates why the task has been difficult. Humans have some innate xenophobic tendencies, and those opposed to free trade are sometimes able to invoke these tendencies to impose restrictions. Analysis of trade issues in military language (trade wars) spreads this incorrect view of exchange. When this is coupled with zero-sum thinking, undesirable policies can easily result. But reduction in trade barriers is an important task for the political system. Recent protests against globalization are another manifestation of tribalism and lack of understanding of the benefits of exchange, which accrue to both sides in a voluntary transaction.

Free international trade has another important benefit. Humans desire peace, and calls for the end of war are common. One suggestion is to create a world government. However, even if such a government could be created, it would itself create serious problems, as the possibility of exit as a check on government power would obviously be lacking. Moreover, since the growth of governments can be explained as a response to balance of power considerations, it is difficult to understand the impetus for a single world government. Increased trade between countries can increase the cost of war, and humans are facultative and can be expected to respond to this increase in cost by reducing conflict. And, increased economic interdependence can increase without bound, and thus is a more reliable method of reducing conflict. Therefore, increased trade and economic interdependence is a more reliable engine for reducing conflict than is increasing the size of government. As long as the possibility of conflict exists, however, we must realize that predation through violence is possible and take appropriate measures for defense.

We can also gain from immigration of skilled foreigners. Such workers can increase our own welfare, as well as their own. We in the United States, which is a desirable target for immigration, could also allow more immigrants who could generate increased wealth for ourselves and for the immigrants. Increased immigration could also solve any problems with our Social Security system (Storesletten 2000). At various times, human xenophobic preferences have been invoked to limit immigration. This is now more difficult in the United States because of our level of ethnic diversity, but it is still an issue to consider in establishing policies.

Affirmative action and other racially based policies are quite dangerous. Humans can easily learn to be ethnocentric and xenophobic. Reducing these tendencies in the United States has been a major achievement and has occurred over the last fifty years. But there is a real danger of rekindling such feelings if policies treat people as members of ethnic groups, rather than as individuals. When government or other entities such as universities use race in decision making, this danger increases.

One of the most important advantages of western society is that women have much more freedom and power in these societies than in any others. For example, Edgerton (1992) points out that wife beating has been approved in virtually every folk society. Indeed, it is this freedom that has allowed feminism to flourish in modern western societies. This is an important benefit, and one that should not be ignored.

One important implication of the analysis is that we should do all we can to restrain the power of government. Public choice scholars have always argued that government employees and politicians are no different from other humans and that there is no strong reason to expect them to act in the public interest if that interest harms their private interests. But the biology strengthens this argument. Humans are strongly selected to maximize fitness, and if this goal conflicts with another goal, we would generally expect fitness to win out. Moreover, individuals differ from each other, and we would expect those who gain the most from using power to achieve their goals to be attracted to government service. Because humans are good at self-deception, we would expect government workers to believe that they are serving the public interest, even as they are acting to maximize their own fitness. Governments have in historic times been major oppressors of humans. In the West, we have crafted a system of government that reduces the ability of government to engage in such oppression. But this does not mean that we can entrust government with unlimited power. Moreover, larger governments encourage more wasteful rent seeking activities, and more free riding. We must maintain vigilance and try to constrain the power of government.

Finally, the most significant policy relevant point of the analysis is the advantage of western, and particularly U.S., society in satisfying human wants. This advantage has been documented above. It exists in many dimensions. We in the West should not be shy in publicizing this advantage and in assisting others to try to emulate our successes.

Notes

Preface

1. Kelly (1995) shows that band sizes of twenty-five are quite common in hunter-gatherer societies. Dunbar (2001) shows that based on brain size, we are adapted to groups of about 150, and we are able to know and interact with this number of individuals. But small bands sometimes coalesce for ritual reasons and exchange of women, so both figures are consistent.

Chapter 1. Background: Evolution and Politics

1. In addition to the references in the Preface, this analysis has been developed in numerous places. The original source is Wilson 1975. See also Dawkins 1976/1989. For major works applying biological reasoning to human behavior, see Alexander 1987; Barkow, Cosmides, and Tooby 1992; Crawford and Krebs 1997; and Miller 2000. There are numerous applications in economics as well; see, for a few examples, Hirshleifer 1977; Rubin and Paul 1979; Rubin and Somanathan 1998; Smith 1992; Robson 1995 and 2001 for a survey; and Siow 1998. Hayek has also written on evolution of preferences, e.g., Hayek 1976. However, his writing was uninformed by more recent advances in evolutionary theory.

2. An exception appears to be the bonobo, the pygmy chimpanzee. This animal seems to have lost much of the aggressiveness of chimpanzees. However, behavior indicates that the species still has some aggressive characteristics and that the ancestral species was very similar to common chimpanzees. See Wrangham and Peterson 1996, chapter 10. Moreover, de Waal 2001 who believes that both chimpanzees and bonobos are relevant for understanding humans, nonetheless indicates that the common ancestor of humans and chimpanzees was more chimpanzee-like in its political behavior, which is relevant for the analysis here.

3. Dolphins may also engage in this behavior: Tooby and Cosmides, in press, 3.

4. Although Darwin had no knowledge of genes; he discovered the process of descent with selection, but did not understand the mechanisms of the process.

Chapter 2. Groups: Membership and Conflict

1. This survey, for 1990–1993, was conducted in forty-three countries.

2. As of 1993, according to *Britannica* online, Hinduism has 13.5 percent, but it is not now a proselytizing religion; no other religion has as much as 10 percent of the world's population. Buddhism has 6 percent. This source indicates that 16.4 percent of the world's population is nonreligious and 4.3 percent are atheists. *Britannica*'s information is consistent with the Inglehart data cited above.

3. This argument was first made by Humphrey 1976. An important authority

who has made this point in numerous publications is Richard Alexander; see, for example, Alexander 1987. Similar arguments are in Bigelow 1969; Byrne and Whiten 1988; and Whiten and Byrne 1997; and, from an economic standpoint, Robson 1995. Dunbar 1996 also argues for a social origin of intelligence.

4. It is possible that climate change and variability in the Pleistocene also had a role in this process, perhaps by leading to larger brains. Consistent with this argument is the fact that brains of many mammalian species increased during this period. However, the sheer magnitude of the increase in human intelligence (both absolutely and relative to all other species) seems to me (and to many others, some cited above) to argue that something more was needed.

5. I have worked as an economist in numerous antitrust issues and have personally seen no evidence that Jews in business compete any less vigorously against other Jews than against gentiles, or than gentiles compete against gentiles. Indeed, one well known paper argued that discrimination against Jews in medical school admissions was to avoid the fact that Jews were well known as highly competitive in pricing (Kessel 1958).

Chapter 3. Altruism, Cooperation, and Sharing

1. Sober and Wilson define groups broadly, so all of the above mechanisms would be forms of group selection in their model.

2. Interestingly, Sober and Wilson (1998, 237) discuss Rawls and the difference principle as an example of a moral rule but do not consider the consistency of this rule with their view of evolved altruism.

3. Actually, the Proposer should offer 1¢, but I assume that 1¢ is worthless.

4. This last is a stock joke phrase; the joke is in the violation of intuitions about ethics.

Chapter 4. Envy

1. *Merriam-Webster's Collegiate Dictionary* (online) provides several definitions of hierarchy. This chapter refers to: "the classification of a group of people according to ability or to economic, social, or professional standing; also: the group so classified." I do not discuss a graded or ranked series, the other sense of hierarchy.

2. In Boehm 1997b, he also argues that this process was a factor in group selection for sociality among humans, but we may ignore this hypothesis for our purposes here.

3. There are of course multiple levels of government—for example, state and municipal governments in the United States. However, it is clear that sovereignty flows from higher to lower level governments; the federal government in the U.S. is clearly more powerful than lower governments.

4. All references are to Marx (1888; web version).

Chapter 5. Political Power

1. The *Merriam-Webster Dictionary* (online) defines history: (a): a chronological record of significant events (as affecting a nation or institution) often including an explanation of their causes; (b): a treatise presenting systematically related natural phenomena. Note that these definitions both require some recording: a chronological record; a treatise.

2. Gordon Tullock, one of the founders of public choice, has been known to say that the best government from his perspective is one with himself as dictator. The next best from the same perspective is a democracy with limited powers, such as we have in the U.S. today. Gordon has a strange sense of humor, but most human males would probably agree with his ranking, suitably modified.

3. Inglehart et al. (1998), table V8. The exception is Northern Ireland.

4. Inglehart et al. (1998), table V10. The exceptions are Latvia (where the figures are 96 percent of the men and 97 percent of the women) and South Korea.

5. There are substantial difficulties in using these numbers to measure inequality. For example, they do not measure the change in position over time; students, for example, are generally in the lower fifth. Nonetheless, even after all adjustments, there will still be substantial inequality in incomes, which however is not translated into inequality in access to females in a monogamous society.

6. These are an alternative explanations to that of Becker and Murphy (1988) that divorce became easier as families began having fewer children.

7. *Statistical Abstract,* table 1412, available: http://www.census.gov/statab/www/

8. *Statistical Abstract,* table 1420, available: http://www.census.gov/statab/www/

9. The gap has increased in absolute terms from two years in 1900 to about six years in 1997: *Statistical Abstract,* table 1421, available: *http://www.census.gov/statab/www/*

10. *Statistical Abstract,* (table 157, available: http://www.census.gov/statab/www/

11. I teach at a private university. For public universities, there are additional hierarchies including a state university system and even the entire state government.

12. The President's Council of Economic Advisers (CEA) as a senior staff economist, the Federal Trade Commission (FTC) as the head of the Division of Consumer Protection within the Bureau of Economics, and the Consumer Product Safety Commission (CPSC) as the Chief Economist.

13. Inglehart et al. (1998) table V5, 83 percent of all respondents viewed family as "very important" (86 percent of females, 80 percent of males.) For work (V4) the corresponding number was 60 percent; for friends (V6), 40 percent; for leisure (V7), 35 percent; for politics (V8), 39 percent; and for religion (V9), 27 percent.

14. I myself took this career track.

Chapter 6. Religion and the Regulation of Behavior

1. An exception is Sen (1970), who does discuss such goods. However, his proof of an impossibility theorem involving such goods has had relatively little impact on economics—certainly less impact than Arrow's Impossibility theorem.

Chapter 7. How Humans Make Political Decisions

1. This is true for rational individual decision making. For decision making by a group, even this axiom could be violated. Kenneth Arrow won the Nobel prize in economics for showing that there is no method of group decision making that can be shown to always lead to consistent decisions (Arrow 1970).

2. Gigerenzer also raises philosophical questions about the meaning of the probability of a single event, but those need not concern us here.

3. I used to think that economists were especially bad at coming up with names for concepts, but apparently psychologists are no better.

4. I suspect that even the most hyperational economist, in moving from one city to another, does not sell all his/her possessions and purchase equal quality used goods at the other end to save moving costs. When I suggested this to my wife before a recent move, to say that she forcefully objected is an understatement.

5. The St. Petersburg Paradox involves a coin-flipping gamble with an infinite expected value. A coin is flipped until a head appears on the nth flip. The payoff is equal to 2^n. The expected value of the game is infinite, but no one will pay a large amount (say, all of their wealth) to play the game. This led the eighteenth-century mathematician Daniel Bernoulli to formulate the theory of risk aversion to replace expected value maximization.

6. *Grimshaw v. Ford Motor Co.*, 119 Cal. App. 3d 757, 174 Cal. Rptr. 348 (1981).

7. I have numerous times attempted to convince product liability defense lawyers that economic testimony about cost-benefit analysis of the sort that is routinely performed at the Consumer Product Safety Commission would help their cases; they have all implicitly agreed with the quoted statement.

Chapter 8. Relevance of the Pleistocene for Today

1. Some believe that increased population and consumption is itself harmful because they believe that the environment cannot sustain this level of consumption (e.g., Ehrlich 1968; Wilson 1998). Others believe that increasing population is associated with increased growth because of technological innovation (e.g., Simon 1981/1986). Since this book deals with human preferences, I will assume that anything leading to increasing human happiness is desirable. However, I do not find the Ehrlich position convincing.

2. I have some difficulties with Frank's analysis. First, he uses flawed (although official) government statistics to argue that real incomes for many have been falling in recent years. Second, his observations are that minor improvements in quality are very expensive. But this is consistent with a situation in which marginal

benefits in quality become very costly to produce, a possible implication of the law of diminishing returns.

3. Turkey has outlawed polygyny, and in many other societies it is not common among educated urban families. However, it may be sufficiently common in rural areas so the class of unmarried males may still be significant.

4. That was one of the most difficult sentences I have ever written. It will undoubtedly cost me many friendships.

References

Abed, Riadh T. 1998. "The Sexual Competition Hypothesis for Eating Disorders." *British Journal of Medical Psychology* 71:525–547.

Adams, Charles Darwin, ed. 1905. *Lysias: Selected Speeches.* Norman: University of Oklahoma Press.

Aiello, Leslie, and Peter Wheeler. 1995. "The Expensive Tissue Hypothesis." *Current Anthropology* 36:199–211.

Alesina, Alberto, Rafael Di Tella, and Robert MacCulloch. 2001. "Inequality and Happiness: Are Europeans and Americans Different?" Working paper, Department of Economics, Harvard University, Cambridge, Mass.

Alexander, Richard D. 1987. *The Biology of Moral Systems.* New York: Aldine de Gruyter.

Alston, Richard M., J. R. Kearl, and Michael B. Vaughn. 1992. "Is There a Consensus Among Economists in the 1990s?" *American Economic Review* 82:203–209.

Ambrose, Stanley H. 2001. "Paleolithic Technology and Human Evolution." *Science* 291:1748.

Appy, Christian. 1993. *Working-Class War: American Combat Soldiers and Vietnam.* Chapel Hill: University of North Carolina Press.

Arkes, Hal R., and Cindy A. Schipani. 1994. "Medical Malpractice v. the Business Judgment Rule: Differences in Hindsight Bias." *Oregon Law Review* 73: 587.

Arnhart, Larry. 1998. *Darwinian Natural Right: The Biological Ethics of Human Nature.* New York: State University of New York Press.

Arrow, Kenneth J. 1970. *Social Choice and Individual Values.* New Haven, Conn.: Yale University Press.

———. 1986. "Agency and the Market." In *Handbook of Mathematical Economics,* Vol. 3, edited by Kenneth J. Arrow and Michael D. Intrilligator. Amsterdam: North Holland.

Axelrod, Robert. 1984. *The Evolution of Cooperation.* New York: Basic Books.

———. 1997. "Promoting Norms," in *The Complexity of Cooperation: Agent-Based Models of Competition and Collaboration.* Princeton, N.J.: Princeton University Press.

Bailey, Martin J. 1992. "Approximate Optimality of Aboriginal Property Rights." *Journal of Law and Economics* 35:183–198.

Barkow, Jerome H., Leda Cosmides, and John Tooby, eds. 1992. *The Adapted Mind.* New York: Oxford University Press.

Barro, Robert J. 1999. "Determinants of Democracy." *Journal of Political Economy* 107, no. 6, part 2:S158–S183.

Baumgartner, Frank R., and Beth L. Leech. 1998. *Basic Interests: The Importance of Groups in Politics and Political Science.* Princeton, N.J.: Princeton University Press.

Becker, Gary S. 1956/1971. *The Economics of Discrimination.* 2d ed. Chicago: University of Chicago Press.

———. 1996. "Norms and the Formation of Preferences," in *Accounting for Tastes*. Cambridge, Mass.: Harvard University Press.

Becker, Gary S., and Kevin M. Murphy. 1988. "The Family and the State." *Journal of Law and Economics* 31: 1–19.

Bentham, Jeremy. 1781/1988. *The Principles of Morals and Legislation*. Amherst, N.Y.: Prometheus Books. Transcribed from: An Introduction to the Principles of Morals and Legislation by Jeremy Bentham—a new edition, corrected by the author. London: Printed for E. Wilson, 1823. General note: "The first edition of this work was printed in the year 1780; and first published in 1789."

Besharov, Douglas J., and Peter Germanis. 2000. "Welfare Reform—Four Years Later." *The Public Interest* 140 (summer): 17–35.

Betzig, Laura L. 1986. *Despotism and Differential Reproduction: A Darwinian View of History*. Hawthorne, N.Y.: Aldine de Gruyter.

———. 1995. "Wanting Women Isn't New—Getting Them Is—Very." *Politics and the Life Sciences* 14: 24–25.

Bigelow, R. S. 1969. *The Dawn Warriors: Man's Evolution Toward Peace*. Boston: Little, Brown.

Bingham, Paul M. 1999. "Human Uniqueness: A General Theory." *Quarterly Review of Biology* 74: 133–169.

Binmore, Ken. 1998. *Just Playing*. Vol. 2 of *Game Theory and the Social Contract*. Cambridge, Mass.: MIT Press.

Bird-David, Nurit. 1992. "Beyond 'The Original Affluent Society': A Culturist Reformulation." *Current Anthropology* 13:33.

Bischof, N. 1978. "On the Phylogeny of Human Morality." In *Morality as a Biological Phenomenon*, edited by Gunther S. Stent. Berkeley: University of California Press.

Boehm, Christopher. 1993. "Egalitarian Behavior and Reverse Dominance Hierarchy." *Current Anthropology* 14:227–254.

———. 1997a. "Egalitarian Behavior and the Evolution of Political Intelligence." In *Machiavellian Intelligence II: Extensions and Evaluations*, edited by Andrew W. Whiten and Richard W. Byrne. Cambridge, U.K.: Cambridge University Press.

———. 1997b. "Impact of the Human Egalitarian Syndrome on Darwinian Selection Mechanics." *American Naturalist* 150: S100–S121. Supplement.

———. 1999. *Hierarchy in the Forest: The Evolution of Egalitarian Behavior*. Cambridge, Mass.: Harvard University Press.

Bowlby, John. 1969. *Attachment*. Vol. 1 of *Attachment and Loss*. New York: Basic Books.

Bowles, Samuel, and Herbert Gintis. 2001. "The Evolution of Strong Reciprocity." Department of Economics. Amherst: University of Massachusetts.

Boyd, Robert, and Peter J. Richerson. 1985. *Culture and the Evolutionary Process*. Chicago: University of Chicago Press.

———. "The Evolution of Indirect Reciprocity." *Social Networks* 11:213–236.

———. 1990. "Group Selection among Alternative Evolutionarily Stable Strategies." *Journal of Theoretical Biology* 145:331–342.

———. 1992. "Punishment Allows the Evolution of Cooperation or Anything Else in Sizeable Groups." *Ethology and Sociobiology* 13:171–195.

Boyer, Pascal 2001. *Religion Explained: The Evolutionary Origins of Religious Thought.* New York: Basic Books.

Bradley, Brenda J. 1999. "Levels of Selection, Altruism and Primate Behavior." *Quarterly Review of Biology* 74: 171–194.

Brown, Donald E. 1991. *Human Universals.* New York: McGraw Hill.

Browne, Kingsley R. 1998. "An Evolutionary Account of Women's Workplace Status." *Managerial and Decision Economics* 19: 427–440.

———. 2001. "Women at War: An Evolutionary Perspective." *Buffalo Law Review* 49, 51–247.

Buchanan, James, and Gordon Tullock. 1965. *The Calculus of Consent.* Ann Arbor: University of Michigan.

Buchanan, James M., Robert D. Tollison, and Gordon Tullock, eds. 1980. *Toward a Theory of the Rent-Seeking Society.* College Station, Tex.: A & M Press.

Burkert, Walter 1996. *Creation of the Sacred: Tracks of Biology in Early Religions.* Cambridge, Mass.: Harvard University Press.

Burnham, Terry, and Jay Phelan. 2000. *Mean Genes.* Cambridge, Mass.: Perseus Publishing.

Buss, David M. 1999. *Evolutionary Psychology: The New Science of the Mind.* Boston: Allyn and Bacon.

Byrne, Richard, and Andrew Whiten, eds. 1988. *Machiavellian Intelligence: Social Expertise and the Evolution of Intelligence in Monkeys. Apes and Humans.* Oxford, U.K.: Clarendon Press.

Calfee, John E., and Clifford Winston. 1993. "The Consumer Welfare Effects of Liability for Pain and Suffering: An Explanatory Analysis." *Brookings Papers on Economic Activity: Microeconomics.* Washington, D.C.: Brookings Institute.

Camerer, Colin. 1995. "Individual Decision Making." In *The Handbook of Experimental Economics,* edited by John H. Kagel and Alvin E. Roth. Princeton, N.J.: Princeton University Press.

Campbell, D. T. 1978. "Social Morality Norms as Evidence of Conflict Between Biological Human Nature and Social System Requirements." In *Morality as a Biological Phenomenon,* edited by Gunter S. Stent. Berkeley: University of California Press.

Caplan, Bryan. 2001. "What Makes People Think Like Economists? Evidence on Economic Cognition from the 'Survey of Americans and Economists on the Economy'" *Journal of Law and Economics* 44, no. 2, 1:395–426.

Caporael, Linda R., and Reuben M. Baron. 1997. "Groups as the Mind's Natural Environment." In *Evolutionary Social Psychology,* edited by Jeffrey A. Simpson and Douglas T. Kenrick. Mahwah, N.J.: Lawrence Erlbaum Associates.

Carmichael, H. Lorne, and W. Bentley Macleod. 1998. "Fair Territory: Preferences, Bargaining, and the Endowment Effect." Working paper, Social Science Research Network Available online: SSRN.com.

Carneiro, Robert L. 2000. "The Transition from Quantity to Quality: A Neglected

Causal Mechanism in Accounting for Social Evolution." *Proceedings of the National Academy of Sciences* 97: 12926–12931.

Carroll, Joseph. 1999. "The Deep Structure of Literary Representations." *Evolution and Human Behavior* 20:159–173.

Cavalli-Sforza, L. Luca, Paolo Menozzi, and Alberto Piazza. 1994. *The History and Geography of Human Genes.* Princeton, N.J.: Princeton University Press.

Charlesworth, William R. 1992. "The Child's Development of the Sense of Justice." In *The Sense of Justice: Biological Foundations of Law,* edited by Roger D. Masters and Margaret Gruter. Newbury Park, Calif.: Sage Publications.

Coase, Ronald H. 1937. "The Nature of the Firm." *Economica* 4: 386–405.

Cosmides, Leda, and John Tooby. 1992. "Cognitive Adaptations for Social Exchange." In *The Adapted Mind,* edited by Jerome H. Barkow, Leda Cosmides, and John Tooby. New York: Oxford University Press.

———. 1996. "Are Humans Good Intuitive Statisticians After All? Rethinking Some Conclusions From the Literature on Judgement Under Uncertainty." *Cognition* 58:1–73.

Crawford, Charles. 1991. "Psychology." In *The Sociobiological Imagination,* edited by Mary Maxwell. Albany, N.Y.: State University of New York Press.

———. 1998. "Environments and Adaptations: Then and Now." In *Handbook of Evolutionary Psychology,* edited by Charles Crawford and Dennis L. Krebs. Mahwah, N.J.: Lawrence Erlbaum Associates.

Crossette, Barbara. 1998. "The World: Democracy's Desert: A Rising Tide of Freedom Bypasses the Arab World." *The New York Times,* April 26.

Curran, Christopher, and Paul H. Rubin. 1995. "The Endowment Effect and Income Transfers. *Research in Law and Economics* 18:225–236.

Daly, Martin, and Margo Wilson. 1988. *Homicide.* New York: Aldine de Gruyter.

Damasio, Antonio. 1994. *Descartes' Error: Emotion, Reason, and the Human Brain.* New York: G. P. Putnam.

Daniels, Norman, Donald W. Light, and Ronald L. Caplan. 1996. *Benchmarks of Fairness for Health Care Reform.* New York: Oxford University Press.

Dawkins, Richard. 1976/1989. *The Selfish Gene.* Oxford: Oxford University Press.

De Waal, Frans B. M. 1982. *Chimpanzee Politics: Power and Sex Among Apes.* London: Jonathan Cape.

———. 1992. "The Chimpanzee's Sense of Social Regularity and Its Relation to the Human Sense of Justice." In *The Sense of Justice: Biological Foundations of Law,* edited by Roger D. Masters and Margaret Gruter. Newbury Park: Sage Publications.

———. 1996. *Good Natured: The Origins of Right and Wrong in Humans and Other Animals.* Cambridge, Mass.: Harvard University Press.

———. 2001. "Apes from Venus: Bonobos and Human Social Evolution." In *Tree of Origin: What Primate Behavior Can Tell Us about Human Social Evolution.* Cambridge, Mass.: Harvard University Press.

Diamond, Jared. 1992. *The Third Chimpanzee: The Evolution and Future of the Human Animal.* New York: HarperCollins.

————. 1998. *Guns, Germs, and Steel.* New York: W. W. Norton.

Donohue, John J. III, and Steven D. Levitt. 2001. "Legalized Abortion and Crime." *Quarterly Journal of Economics* 116:379–420.

Dugger, Celia W. 2001. "Modern Asia's Anomaly: The Girls Who Don't Get Born." *New York Times,* May 6.

Dunbar, Robin. 1996. *Grooming, Gossip, and the Evolution of Language.* Cambridge, Mass.: Harvard University Press.

————. 1998. "The Social Brain Hypothesis." *Evolutionary Anthropology* 6:178–190.

————. 2001. "Brains on Two Legs: Group Size and the Evolution of Intelligence." In *Tree of Origin: What Primate Behavior Can Tell Us about Human Social Evolution,* edited by Frans B. M. de Waal. Cambridge, Mass.: Harvard University Press.

Durham, William H. 1991. *Coevolution: Genes, Cultures and Human Diversity.* Stanford, Calif.: Stanford University Press.

Edgerton, Robert B. 1992. *Sick Societies: Challenging the Myth of Primitive Harmony.* New York: Free Press.

Ehrlich, Paul. 1968. *The Population Bomb.* New York: Ballentine.

Eller, Cynthia. 2000. *The Myth of Matriarchal Prehistory.* Boston: Beacon Press.

Ellingsen, Tore. 1997. "The Evolution of Bargaining Behavior." *Quarterly Journal of Economics* 112:581–602.

Epstein, Richard A. 1980. "A Taste for Privacy: The Evolution of a Naturalistic Ethic." *Journal of Legal Studies* 9:655.

————. 1989. "The Utilitarian Foundations of Natural Law." *Harvard Journal of Law and Public Policy* 12:713.

————. 1990. "The Varieties of Self-Interest." *Social Philosophy and Policy* 8:102.

Evans, Jonathan St. B. T., and David E. Over. 1996. *Rationality and Reasoning.* East Sussex, U.K.: Psychology Press.

Fehr, Ernst, and Klaus M. Schmidt. 2002. "Theories of Fairness and Reciprocity— Evidence and Economic Applications." In *Advances in Economics and Econometrics,* 8th World Congress, edited by M. Dewatripont, L. Hansen, and St. Turnovsky. Cambridge: Cambridge University Press.

Fischel, Daniel R. 1995. *Payback: The Conspiracy to Destroy Michael Milken and His Financial Revolution.* New York: HarperBusiness.

Foley, Robert. 1995. *Humans Before Humanity.* Cambridge, Mass.: Blackwell.

Frank, Robert. 1988. *Passions Within Reasons: The Strategic Control of the Emotions.* New York: W. W. Norton.

————. 2000. *Luxury Fever.* Princeton, N.J.: Princeton University Press.

Geary, David C. 1998. *Male, Female: The Evolution of Human Sex Differences.* Washington, D.C.: American Psychological Association.

Ghiglieri, Michael P. 1999. *The Dark Side of Man: Tracing the Origins of Male Violence.* Reading, Mass.: Perseus Books.

Gigerenzer, Gerd. 1991. "How to Make Cognitive Illusions Disappear: Beyond Heuristics and Biases." *European Review of Social Psychology* 2: 83–115.

————. 1996. "On Narrow Norms and Vague Heuristics: A Reply to Kahneman and Tversky, 1996." *Psychological Review* 100:592–596.

Gigerenzer, Gerd, Peter M. Todd, and the ABC (Center for Adaptive Behavior and Cognition) Research Group. 1999. *Simple Heuristics That Make Us Smart.* New York: Oxford University Press.

Gilman, Sander L. 1999. *Making the Body Beautiful: A Cultural History of Aesthetic Surgery.* Princeton, N.J.: Princeton University Press.

Gintis, Herbert. 2000a. *Game Theory Evolving.* Princeton, N.J.: Princeton University Press.

———. 2000b. "Strong Reciprocity and Human Sociality." *Journal of Theoretical Biology* 211: 169–179.

Glassner, Barry. 1999. *The Culture of Fear.* New York: Basic Books.

Glover, Jonathan. 2000. *Humanity: A Moral History of the Twentieth Century.* New Haven, Conn.: Yale University Press.

Goetze, David. 1998. "Evolution, Mobility, and Ethnic Group Formation." *Politics and the Life Sciences* 17, 1:59–71.

Goodall, Jane. 1986. *The Chimpanzees of Gombe.* Cambridge, Mass.: Harvard University Press.

Gowlett, John A. J. 1992. "Tools—the Paleolithic Record" In *The Cambridge Encyclopedia of Human Evolution, edited by Steven Jones, Robert Martin, and David Pilbeam.* Cambridge, U.K.: Cambridge University Press.

Grady, Mark F., and Michael T. McGuire. 1997. "A Theory of the Origin of Natural Law": Law, Human Behavior and Evolution (Symposium issue, edited by Gail L. Herriot). *Journal of Contemporary Legal Issues,* Symposium Issue: 8:87–129.

———. 1999. "The Nature of Constitutions." *Journal of Bioeconomics* 1:227–240.

Greif, Avner. 1993. "Contract Enforceability and Economic Institutions in Early Trade: The Maghribi Traders' Coalition." *American Economic Review* 83:525–548.

Guthrie, Stewart. 1993. *Faces in the Clouds: A New Theory of Religion.* New York: Oxford University Press.

Hanson, Jon D., and Douglas A. Kysar. 1999. "Taking Behavioralism Seriously: Some Evidence of Market Manipulation." *Harvard Law Review* 112:1422–1572.

Hansson, Ingemar, and Charles Stuart. 1990. "Malthusian Selection of Preferences." *American Economic Review* 80:529–544.

Harden, Blaine 1999. "Stresses of Milosevic's Rule Blamed for Decline in Births." *The New York Times,* July 5.

Hayek, Friedrich A. von. 1976. *The Mirage of Social Justice.* Vol 2 of *Law, Legislation and Liberty.* Chicago: University of Chicago Press.

———. 1944. *The Road to Serfdom.* Chicago: University of Chicago Press.

Henrich, Joseph, Robert Boyd, Samuel Bowles, Colin Camerer, Ernst Fehr, Herbert Gintis, and Richard McElreath. 2001. "In Search of Homo Economicus: Behavioral Experiments in 15 Small-Scale Societies." *American Economic Review, Papers and Proceedings* 91:73–78.

Hinde, Robert A. 1999. *Why Gods Persist: A Scientific Approach to Religion.* New York: Routledge.

Hirshleifer, Jack. 1977. "Economics from a Biological Viewpoint." *Journal of Law and Economics* 20:1–52.

——. 1980. "Privacy: Its Origin, Function and Future." *Journal of Legal Studies* 9:194–210.

——. 1998. "The Bioeconomic Causes of War": Management. Organization and Human Nature (Special issue, edited by Livia Markoczy). *Managerial and Decision Economics* 19:457–466.

——. 1999. "There Are Many Evolutionary Pathways to Cooperation." *Journal of Bioeconomics* 1:73–93.

Hirshleifer, Jack, and J. Martinez-Coll. 1988. "What Strategies Can Support the Evolutionary Emergence of Cooperation?" *Journal of Conflict Resolution* 32:367–398.

Hochman, Harold, and James D. Rodgers. 1969. "Pareto Optimal Redistribution." *American Economic Review* 59:542–557.

Hodges, S. Blair. 2000. "Human Evolution: A Start for Population Genomics." *Nature* 408: 652.

Hoffman, Elizabeth, Kevin A. McCabe, and Vernon L. Smith. 1998. "Behavioral Foundations of Reciprocity: Experimental Economics and Evolutionary Psychology." *Economic Inquiry* 36:335–352.

Hrdy, Sarah Blaffer. 1999. *Mother Nature: A History of Mothers, Infants, and Natural Selection.* New York: Pantheon.

Huck, Steffen, Joerg Oechssler, and Georg Kirchsteiger. 1999. "Learning to Like What You Have—Explaining the Endowment Effect." Working paper, Humboldt University, Berlin.

Humphrey, Nicholas. 1976. "The Social Function of the Intellect." In *Growing Points in Ethology,* edited by P.P.G. Bateson. Cambridge, U.K.: Cambridge University Press.

Iannaccone, Laurence R. 1998. "Introduction to the Economics of Religion." *Journal of Economic Literature* 36:1465–1496.

Inglehart, Ronald, Miguel Basañez, and Alejandro Moreno. 1998. *Human Values and Beliefs: A Cross-Cultural Sourcebook; Political, Religious, Sexual and Economic Norms in 43 Countries: Findings from the 1990–1993 World Values Survey.* Ann Arbor: University of Michigan Press.

Irons, William 1998. "Adaptively Relevant Environments Versus the Environment of Evolutionary Adaptedness." *Evolutionary Anthropology* 6: 194–204.

Janicki, Maria, and Dennis L. Kreps. 1998. "Evolutionary Approaches to Culture." In *Handbook of Evolutionary Psychology,* edited by Charles Crawford and Dennis L. Krebs. Mahwah, N.J.: Lawrence Erlbaum Associates.

Jenni, Karen E., and George Loewenstein. 1997. "Explaining the 'Identifiable Victim' Effect." *Journal of Risk and Uncertainty* 14:235–257.

Johnson, Gary R. 1986. "Kin Selection, Socialization, and Patriotism: An Integrating Theory." *Politics and the Life Sciences* 4:127–140.

Johnson, Victor S. 1999. *Why We Feel: The Science of Human Emotions.* Reading. Mass.: Perseus Books.

Jones, Charles I. 2001. "Was an Industrial Revolution Inevitable? Economic Growth Over the Very Long Run." *Advances in Macroeconomics* 1 (2).

Jones, Steven, Robert Martin, and David Pilbeam 1992. *The Cambridge Encyclopedia of Human Evolution.* Cambridge, U.K.: Cambridge University Press.

Kahneman, Daniel, and Amos Tversky. 1996. "On the Reality of Cognitive Illusions." *Psychological Review* 103: 582–591.

———. 1979. "Prospect Theory: An Analysis of Decision Under Risk." *Econometrica* 47:263–291.

Kahneman, Daniel, Jack L. Knetsch, and Richard H. Thaler. 1990. "Experimental Effects of the Endowment Effect and the Coase Theorem." *Journal of Political Economy* 98: 1325–1348.

Kahneman, Daniel, Paul Slovic, and Amos Tversky. 1982. *Judgement Under Uncertainty: Heuristics and Biases.* Cambridge, U.K.: Cambridge University Press.

Kamin, Kim A., and Jeffrey J. Rachlinski. 1995. "Ex Post v Ex Ante: Determining Liability in Hindsight." *Law and Human Behavior* 19: 89.

Kaplan, Hillard, Kim Hill, Jane Lancaster, and A. Magdelana Hurtado. 2000. "A Theory of Human Life History Evolution: Diet, Intelligence, and Longevity." *Evolutionary Anthropology* 9:156–185.

Kau, James B., and Paul H. Rubin. 1979. "Self Interest, Ideology and Logrolling in Congressional Voting." *Journal of Law and Economics* 22:365–384.

Kearl, James R., Clayne L. Pope, Gordon T. Whiting, and Larry Wimmer. 1979. "What Economists Think." *American Economic Review* 69:28–37.

Keeley, Lawrence H. 1996. *War Before Civilization.* New York: Oxford University Press.

Keenan, Donald. 1981. "Value Maximization and Welfare Theory." *Journal of Legal Studies* 10: 409–419.

Kelly, Robert L. 1995. *The Foraging Spectrum: Diversity in Hunter-Gatherer Lifeways.* Washington. D.C.: Smithsonian Institution Press.

Kessel, Reuben A. 1958. "Price Discrimination in Medicine." *Journal of Law and Economics.* (October):22–53.

Klein, Benjamin, and Keith Leffler. 1981. "The Role of Market Forces in Assuring Contractual Performance." *Journal of Political Economy* 89: 615–641.

Klein, Benjamin, Robert Crawford, and Armen Alchian. 1978. "Vertical Integration, Appropriable Rents, and the Competitive Contracting Process." *Journal of Law and Economics* 21:297.

Klein, Richard G. 2000. "Archeology and the Evolution of Human Behavior." *Evolutionary Anthropology* 9:17–36.

Knauft, Bruce. 1991. "Violence and Sociality in Human Evolution." *Current Anthropology* 32:391–428.

———. 2000. "Symbols. Sex and Sociality in the Evolution of Human Morality." In *Evolutionary Origins of Human Morality,* edited by Leonard D. Katz. Exeter, U.K.: Imprint Academic.

Krebs, Dennis L. 1998. "The Evolution of Moral Behaviors." In *Handbook of Evolutionary Psychology,* edited by Charles Crawford and Dennis L. Krebs. Mahwah, N.J.: Lawrence Erlbaum Associates.

Krebs, Dennis L., and Kathy Denton. 1997. "Social Illusions and Self-Deception:

The Evolution of Biases in Person Perception." In *Evolutionary Social Psychology*, edited by Jeffry A. Simpson and Douglas T. Kenrick. Mahwah, N.J.: Lawrence Erlbaum Associates.

Kremer, Michael. 1993. "Population Growth and Technological Change: One Million B.C to 1990." *Quarterly Journal of Economics* 108:681–716.

Kristof, Nicholas D., and Sheryl WuDunn. 2000. "Two Cheers for Sweatshops." *The New York Times*, September 24.

Kronholz, June. 1999. "School Firearm Expulsions Dropped in '97–'98." *The Wall Street Journal*, August 11.

Kronk, Lee. 1999. *That Complex Whole: Culture and the Evolution of Human Behavior*. Boulder, Colo.: Westview Press.

Krugman, Paul. 1997. *Pop Internationalism*. Cambridge, Mass.: MIT Press.

———. 2001. "Hearts and Heads." *New York Times*, April 22.

Kuran, Timur. 1998. "Ethnic Norms and their Transformation Through Reputational Cascades." *Journal of Legal Studies* 27:623–659.

Kurzban, R. O. 1999. *The Social Psychophysics of Cooperation in Groups*. Ph.D. dissertation, University of California, Santa Barbara. Quoted in Pascal Boyer. 2001. *Religion Explained: The Evolutionary Origins of Religious Thought* (New York: Basic Books).

Lakoff, George. 1996. *Moral Politics: What Conservatives Know that Liberals Don't*. Chicago: University of Chicago Press.

Landa, Janet. 1981. "A Theory of the Ethnically Homogeneous Middleman Group: An Institutional Alternative to Contract Law." *Journal of Legal Studies* 10:349–362.

La Porta, Rafael, Florencio Lopez-de-Silanes, Andrei Shleifer, and Robert Vishny. 1999. "The Quality of Government." *Journal of Law, Economics and Organization* 15:222–279.

Ledyard, John O. 1995. "Public Goods: A Survey of Experimental Research." In *Handbook of Experimental Economics*, edited by John H. Kagel and Alvin E. Roth. Princeton, N.J.: Princeton University Press.

Lelyveld, Joseph. 2001. "All Suicide Bombers Are Not Alike." *New York Times Magazine*, October 28, late edition, final, section 6, 49.

Lott, John R., Jr., and Larry Kenny. 1999. "Did Women's Suffrage Change the Size and Scope of Government?" *Journal of Political Economy* 107:1163–1198.

Low, Bobbi S. 2000. *Why Sex Matters: A Darwinian Look at Human Behavior*. Princeton, N.J.: Princeton University Press.

MacDonald, Kevin. 1994 *A People That Shall Dwell Alone: Judaism As a Group Evolutionary Strategy*. Westport, Conn.: Praeger.

———. 1995. "The Establishment and Maintenance of Socially Imposed Monogamy in Western Europe." *Politics and the Life Sciences* 14:2–23.

———. 1998a. *Separation and Its Discontents: Toward an Evolutionary Theory of Anti-Semitism*. Westport, Conn.: Praeger.

———. 1998b. *The Culture of Critique*. Westport, Conn.: Praeger.

Manson, Joseph H., and Richard W. Wrangham. 1991. "Intergroup Aggression in Chimpanzees and Humans." *Current Anthropology* 32: 369–390.

Marshall, Lorna 1961. "Sharing, Talking and Giving: Relief of Social Tensions among !Kung Bushmen." *Africa* 31:231–249.

Marx, Karl. 1888. *Manifesto of the Communist Party.* Web edition: Project Guttenberg, available online: http://www.thalasson.com/gtn/gtnletM.htm#marxkarl

Maryanski, Alexandra, and Jonathan H. Turner. 1992. *The Social Cage: Human Nature and the Evolution of Society.* Stanford, Calif.: Stanford University Press.

Masters, Roger D. 1989. *The Nature of Politics.* New Haven, Conn.: Yale University Press.

Maynard Smith, John. 1982. *Evolution and the Theory of Games.* New York: Cambridge University Press.

Maynard Smith, John, and Eörs Szathmáry, eds. 1999. *The Origins of Life: From the Birth of Life to the Origins of Language.* New York: Oxford University Press.

Mayr, Ernst. 1988. "The Probability of Extraterrestrial Intelligent Life." In *Toward a New Philosophy of Biology.* Cambridge, Mass.: Belknap Press of Harvard University Press.

McCaffery, Edward J., Daniel J. Kahneman, and Matthew L. Spitzer. 1995. "Framing the Jury: Cognitive Perspectives on Pain and Suffering Awards." *Virginia Law Review* 81:1341.

McGee, John S. 1958. "Predatory Price Cutting: The Standard Oil N.J. Case." *Journal of Law and Economics* 1: 137–169.

McGinnis, John O. 1997. "The Human Condition and Constitutional Law: A Prolegomenon." *Journal of Contemporary Legal Issues,* Symposium Issue: *Law, Human Behavior and Evolution* 8:211–240.

McGuire, Michael T. 1992. "Moralistic Aggression, Processing Mechanisms, and the Brain: The Biological Foundations of the Sense of Justice." In *The Sense of Justice: Biological Foundations of Law,* edited by Roger D. Masters and Margaret Gruter. Newbury Park, Calif.: Sage Publications.

Mealey, Linda. 1995. "The Sociobiology of Sociopathy: An Integrated Evolutionary Model." *Behavioral and Brain Sciences* 18:523–599.

Mesquida, Christian G., and Neil I. Weiner. 1996. "Human Collective Aggression: A Behavioral Ecology Perspective." *Ethology and Sociobiology* 17:247–262.

Miller, Alan S., and John P. Hoffman. 1995. "Risk and Religion: An Explanation of Gender Differences in Religion." *Journal for the Scientific Study of Religion* 34: 63–75.

Miller, Geoffrey F. 1997. "Protean primates: The evolution of adaptive unpredictability in competition and courtship." In *Machiavellian Intelligence II: Extensions and Evaluations,* edited by Andrew Whiten and Richard W. Byrne. Cambridge, U.K.: Cambridge University Press.

———. 2000. *The Mating Mind: How Sexual Choice Shaped the Evolution of Human Nature.* New York: Doubleday.

Moore, Randall F. 1996. "Caring for Identified Versus Statistical Lives: An Evolutionary View of Medical Distributive Justice." *Ethology and Sociobiology* 17: 379–401.

Mullen, Brian, Rupert Brown, and Colleen Smith. 1992. "Ingroup Bias as a Function of Salience, Relevance, and Status: An Integration." *European Journal of Social Psychology* 22: 103–122.

Nelson, Phillip. 1994. "Voting and Imitative Behavior." *Economic Inquiry* 32:92–102.

Nozick, Robert. 1974. *Anarchy. State. and Utopia.* New York: Basic Books.

Olson, James M., Philip A. Vernon, and Julie Aitken Harris. 2001. "The Heritability of Attitudes: A Study of Twins." *Journal of Personality and Social Psychology* 80:845.

Olson, Mancur. 1965. *The Logic of Collective Action.* Cambridge, Mass.: Harvard University Press.

Ostrom, Eleanor. 2000. "Collective Action and the Evolution of Social Norms." *Journal of Economic Perspectives* 14:137–158.

Peltzman, Sam. 1973. "An Evaluation of Consumer Protection Legislation: The 1962 Drug Amendments." *Journal of Political Economy* 81:1049–1091.

Philipson, Tomas J., and Richard A. Posner. 1999. "The Long-Run Growth in Obesity as a Function of Technological Change." Law and Economics working paper 78, University of Chicago Law School.

Pinker, Steven. 1994. *The Language Instinct.* New York: W. Morrow.

———. 1997. *How the Mind Works.* New York: W. W. Norton.

Pinker, Steven, and Paul Bloom. 1992. "Natural Language and Natural Selection." In *The Adapted Mind,* edited by Jerome H. Barkow, Leda Cosmides, and John Tooby. New York: Oxford University Press.

Pipes, Daniel. 2001. "The Danger Within: Militant Islam in America." *Commentary* 112, 4:19–24.

Pipes, Richard. 1990. *The Russian Revolution.* New York: Vintage Books.

———. 1999. *Property and Freedom.* New York: Knopf.

Poole, Keith T., and Howard Rosenthal. 1997. *Congress: A Political-Economic History of Roll Call Voting.* New York: Oxford University Press.

Posner, Richard A. 1979. "Utilitarianism, Economics, and Legal Theory." *Journal of Legal Studies* 8:103.

———. 1980. "A Theory of Primitive Society, With Special Reference to Law." *Journal of Law and Economics* 23:1–53.

———. 1992. *Sex and Reason.* Cambridge, Mass.: Harvard University Press.

———. 1999. *The Problematics of Moral and Legal Theory.* Cambridge, Mass.: Belknap Press of Harvard University Press.

Potts, Richard. 1998. "Variability Selection in Hominid Evolution." *Evolutionary Anthropology* 7:81–95.

Powell, Irene. 1987. "The Effect of Reductions in Concentration on Income Distribution." *The Review of Economics and Statistics* 69:75–82.

Premack, David, and Ann James Premack. 1994. "Moral Belief: Form versus Content." In *Mapping the Mind,* edited by Lawrence A. Hirschfeld and Susan A. Gelman. Cambridge, N.Y.: Cambridge University Press.

Pusey, Anne E. 2001. "Of Genes and Apes: Chimpanzee Social Organization and Reproduction." In *Tree of Origin: What Primate Behavior Can Tell Us about Human*

Social Evolution, edited by Frans B. M. de Waal. Cambridge, Mass.: Harvard University Press.

Rabin, Matthew. 1998. "Psychology and Economics." *Journal of Economic Literature* 36:11–47.

Rawls, John. 1971. *A Theory of Justice.* Cambridge, Mass.: Belknap Press of Harvard University Press.

Reiss, Steven. 2000. *Who Am I? The 16 Basic Desires That Motivate Our Actions and Define Our Personalities.* New York: Tarcher/Putnam.

Richerson, Peter J., and Robert Boyd. 1999. "Complex Societies: The Evolutionary Origins of a Crude Superorganism": Special issue on Group Selection, edited by Christopher Boehm. *Human Nature* 10, 3:253–289.

Ridley, Matt. 1996. *The Origins of Virtue: Human Instincts and the Evolution of Cooperation.* New York: Viking Press.

Robson, Arthur J. 1995. "The Evolution of Strategic Behavior." *Canadian Journal of Economics* 28:17–41.

———. 2001. "The Biological Basis of Economic Behavior." *Journal of Economic Literature* 39:11–33.

Rode, Catrin, and X. T. Wang. 2000. "Risk-Sensitive Decision Making Examined Within an Evolutionary Framework." *American Behavioral Scientist* 43:926–939.

Rode, Catrin, Leda Cosmides, Wolfgang Hell, and John Tooby. 1999. "When and why do people avoid unknown probabilities in decisions under uncertainty? Testing some predictions from optimal foraging theory." *Cognition* 72:269–304.

Roe, Mark J. 1998. "Backlash." *Columbia Law Review* 98:217–241.

Roes, Frans L. 1995. "The Size of Societies, Stratification, and Belief in High Gods Supportive of Human Morality." *Politics and the Life Sciences* 14:73–77.

Rogers, Alan. 1994. "Evolution of Time Preference by Natural Selection." *American Economic Review* 84:460–481.

Rose, Michael R. 1998. *Darwin's Spectre: Evolutionary Biology in the Modern World.* Princeton, N.J.: Princeton University Press.

Roseberry, William. 1988. "Political Economy." *Annual Review of Anthropology* 17:161–185.

Roth, Alvin E. 1995. "Introduction to Experimental Economics." In *The Handbook of Experimental Economics,* edited by John H. Kagel and Alvin E. Roth. Princeton, N.J.: Princeton University Press.

Rubin, Paul H. 1973 "The Economic Theory of the Criminal Firm." In *The Economics of Crime and Punishment,* edited by Simon Rottenberg. Washington. D.C.: American Enterprise Institute.

———. 1977. "Why is the Common Law Efficient." *Journal of Legal Studies* 6:51–63.

———. 1982. "Evolved Ethics and Efficient Ethics." *Journal of Economic Behavior and Organization* 3:161–74.

———. 1990. *Managing Business Transactions: Controlling the Cost of Coordinating. Communicating and Decision Making.* New York: Free Press.

———. 1993. *Tort Reform by Contract.* Washington, D.C.: American Enterprise Institute.

———. 1994. "Growing A Legal System in the Post-Communist Economies." *Cornell International Law Journal* 27:1–47.

Rubin, Paul H., and Chris Paul. 1979. "An Evolutionary Model of Taste for Risk." *Economic Inquiry* 17:585–596.

Rubin, Paul H., and E. Somanathan. 1998. "Humans as Factors of Production: An Evolutionary Analysis": Management. Organization and Human Nature (Special issue, edited by Livia Markoczy). *Managerial and Decision Economics* 19, nos. 7–8, 441–445.

Rubin, Paul H., Christopher Curran, and John Curran. 2001. "Litigation versus Lobbying: Forum Shopping by Rent-Seekers." *Public Choice* 107:295–310.

Rubin, Paul H., James B. Kau, and Edward Meeker. 1979. "Forms of Wealth and Parent-Offspring Conflict." *Journal of Social and Biological Structures* 2:53–64.

Rummel, R. J. 1994. *Death by Government*. New Brunswick, N.J.: Transaction Publishers.

Saffer, Henry, and Frank Chaloupka. 1999. "The Demand for Illicit Drugs." *Economic Inquiry* 37:401–411.

Sahlins, Marshall. 1972. *Stone Age Economics*. Chicago: Aldine-Atherton.

Samuelson, Larry. 2001. "Analogies, Adaptation, and Anomalies." *Journal of Economic Theory* 97:320.

Scalise Sugiyama, Michelle. 2001. "Food, Foragers, and Folklore: The Role of Narrative in Human Subsistence." *Evolution and Human Behavior* 22:221.

Scherer, F. M., and David Ross. 1990. *Industrial Market Structure and Economic Performance*. 3rd ed. Boston: Houghton Mifflin.

Schumpeter, Joseph. 1942/1950. *Capitalism, Socialism, Democracy*. New York: Harper.

Schwartz, Gary T. 1991. "The Myth of the Ford Pinto Case." *Rutgers Law Review* 43:1013.

Scully, Gerald. 1997. "Democide and Genocide as Rent-Seeking Activities." *Public Choice* 33:71–97.

Sen, Amartya. 1970. "The Impossibility of a Paretian Liberal." *Journal of Political Economy* 78:152–157.

———. 1999. "The Possibility of Social Choice." *American Economic Review* 89:349–378.

Sethi, Rajiv, and E. Somanathan. 1996. "The Evolution of Social Norms in Common Property Resource Use." *American Economic Review* 86:766–788.

Shaw, R. Paul, and Yuwa Wong. 1989. *Genetic Seeds of Warfare: Evolution, Nationalism, and Patriotism*. Boston: Unwin Hyman.

Shepher, J. 1971. "Mate Selection Among Second-Generation Kibbutz Adolescents and Adults: Incest Avoidance and Negative Imprinting." *Archives of Sexual Behavior* 1:293–307.

Shughart, William F. II. 1997. *The Organization of Industry*. Houston, Tex.: DAME Publications.

Sidanius, Jim, Felicia Pratto, and Michael Mitchell. 1994. "In-group Identification, Social Dominance Orientation, and Differential Intergroup Social Allocation." *Journal of Social Psychology* 134:151.

Silverman, Irwin, and Krista Phillips. 1998. "The Evolutionary Psychology of Spatial Sex Differences." In *Handbook of Evolutionary Psychology*, edited by Charles Crawford and Dennis L. Krebs. Mahwah, N.J.: Lawrence Erlbaum Associates.

Simon, Herbert A. 1997. *Models of Bounded Rationality: Empirically Grounded Economic Reason*. Cambridge: MIT Press.

Simon, Julian. 1981/1996. *The Ultimate Resource*. Princeton, N.J.: Princeton University Press.

——. 1999. *Hoodwinking the Nation*. New Brunswick, N.J.: Transaction Publishers.

Singer, Max. 1999. "The Population Surprise." *Atlantic Monthly* August. Available online: www.theatlantic.com

Singer, Peter. 2000. *A Darwinian Left: Politics, Evolution, and Cooperation*. New Haven, Conn.: Yale University Press.

Siow, Aloysius. 1998. "Differential Fecundity, Markets, and Gender Roles." *Journal of Political Economy* 106:334–354.

Siwolop, Sana. 2000. "Nooses, Symbols of Race Hatred, at Center of Workplace Lawsuits." *New York Times*, July 10.

Skyrms, Brian. 1996. *Evolution of the Social Contract*. New York: Cambridge University Press.

Slovic, Paul. 2000. *The Perception of Risk*. Sterling, Va.: Earthscan.

Smith, Adam. 1789/1994. *An Inquiry into the Nature and Causes of the Wealth of Nations*, edited by Edwin Cannan. New York: The Modern Library.

Smith, Bruce D. 1995. *The Emergence of Agriculture*. New York: Scientific American Library.

Smith, Vernon L. 1991. "Economic Principles in the Emergence of Humankind." *Economic Inquiry* 29:1–13.

——. 1998. "The Two Faces of Adam Smith." *Southern Economic Journal* 65:1–19.

Sober, Elliott, and David Sloan Wilson. 1998. *Unto Others: The Evolution and Psychology of Unselfish Behavior*. Cambridge, Mass.: Harvard University Press.

Sowell, Thomas 1983. *The Economics and Politics of Race: An International Perspective*. New York: Quill.

——. 1990. *Preferential Policies: An International Perspective*. New York: W. Morrow.

Spence, Michael. 1973. "Job Market Signaling." *Quarterly Journal of Economics* 87:355–374.

Stanford, Craig B. 1999. *The Hunting Apes: Meat Eating and the Origins of Human Behavior*. Princeton, N.J.: Princeton University Press.

Stanford, Craig B., and John S. Allen. 1991. "On Strategic Storytelling: Current Models of Human Behavioral Evolution." *Current Anthropology* 32:58–61.

Stanley, Thomas J., and William D. Danko. 1998. *The Millionaire Next Door: The Surprising Secrets of America's Wealthy*. New York: Pocket Books.

Stark, Rodney. 1997. *The Rise of Christianity*. Princeton, N.J.: Princeton University Press.

Stein, Benjamin. 1979. *The View From Sunset Boulevard*. New York: Basic Books.

Stiner, Mary C., Natalie D. Munro, Todd A. Surovell, Eitan Tchernov, and Ofer Bar-Yosef. 1998. "Paleolithic Population Growth Pulses Evidenced by Small

Animal Exploitation." *Science* September 25. Available online: hppt://www
.sciencemag.org.

Storesletten, Kjetil. 2000. "Sustaining Fiscal Policy through Immigration." *Journal of Political Economy* 108:300.

Sunstein, Cass R. 1993. *The Partial Constitution*. Cambridge, Mass.: Harvard University Press.

Tajfel, H. 1970. "Experiments in Intergroup Discrimination." *Scientific American* 223, 11:96–102.

Tesser, Abraham 1993. "The Importance of Heritability in Psychological Research: The Case of Attitudes." *Psychological Review* 100:129–142.

Thaler, Richard H. 1992. *The Winner's Curse: Paradoxes and Anomalies of Economic Life*. New York: Free Press.

Tiger, Lionel. 1969. *Men in Groups*. New York: Random House.

Tollision, Robert. 1982. "Rent Seeking: A Survey." *Kylkos* 35:575–601.

Tooby, John, and Leda Cosmides. 1996. "Friendship and the Banker's Paradox: Other Pathways to the Evolution of Adaptations for Altruism": Evolution of Social Behaviour Patterns in Primates and Man (Special issue, edited by W. G. Runciman, John Maynard Smith, and R. I. M. Dunbar). *Proceedings of the British Academy* 88:119–143.

———. In press. The Evolution of Coalitional Aggression and Its Cognitive Foundations. In *Evolutionary Psychology: Foundational Papers*. Cambridge, Mass.: MIT Press.

Trivers, Robert L. 1971. "The Evolution of Reciprocal Altruism." *Quarterly Review of Biology* 46:35–57.

———. 1972. "Parental Investment and Sexual Selection." In *Sexual Selection and the Descent of Man 1871–1971*, edited by Bernard Campbell. Chicago: Aldine Publishing Company.

Tudge, Colin. 1998. *Neanderthals, Bandits, and Farmers: How Agriculture Really Began*. New Haven, Conn.: Yale University Press.

Tullock, Gordon. 1967. "The Welfare Costs of Tariffs, Monopolies, and Theft." *Western Economic Journal* 5:224–232.

———. 1999. "Some Personal Reflections on the History of Bioeconomics." *Journal of Bioeconomics* 1:13–18.

United Nations. 2001. *Human Development Report 2001*. New York: Oxford University Press. Available online: http://www.undp.org/hdr2001

Wax, Amy L. 2000. "Rethinking Welfare Rights: Reciprocity Norms, Reactive Attitudes, and the Political Economy of Welfare Reform." *Law and Contemporary Problems* 63:257.

Weisfeld, Glenn E. 1990. "Sociobiological Patterns of Arab Culture." *Ethology and Sociobiology* 11:23–49.

Westermarck, Edward. 1932. *Ethical Relativity*. New York: Harcourt, Brace and Company.

Whiten, Andrew, and Richard W. Byrne, eds. 1997. *Machiavellian Intelligence II: Extensions and Evaluations*. Cambridge, U.K.: Cambridge University Press.

Williams, George C. 1966. *Adaptation and Natural Selection: A Critique of Some Current Evolutionary Thought.* Princeton, N.J.: Princeton University Press.

Williamson, Oliver. 1983. *Markets and Hierarchies: Analysis and Antitrust Implications.* New York: Free Press.

Wills, Christopher. 1998. *Children of Prometheus: The Accelerating Pace of Human Evolution.* Reading, Mass.: Perseus Books.

Wilson, David Sloan. 1999. "A Critique of R. D. Alexander's Views on Group Selection." *Biology and Philosophy* 14:431–499.

Wilson, Edward O. 1975. *Sociobiology: The New Synthesis.* Cambridge, Mass.: Harvard University Press.

———. 1998. *Consilience: The Unity of Knowledge.* New York: Knopf.

Wilson, J. Q. 1993. *The Moral Sense.* New York: Free Press.

Winslow, Ron. 1999. "Violence Among Teenagers Appears to be Waning." *The Wall Street Journal,* August 4.

Wittfogel, Karl A. 1957. *Oriental Despotism: A Comparative Study of Total Power.* New Haven, Conn.: Yale University Press.

Wolfe, Alan. 2000. "The Moral Sense in Estate Tax Repeal." *New York Times,* July 24.

Wolpoff, Milford, and Rachel Gaspari. 1997. *Race and Human Evolution: A Fatal Attraction.* Boulder, Colo.: Westview Press.

Wrangham, Richard W. 2001. "Out of the *Pan* into the Fire: How Our Ancestors Evolution Depended on What they Ate." In *Tree of Origin: What Primate Behavior Can Tell Us about Human Social Evolution,* edited by Frans B. M. de Waal. Cambridge, Mass.: Harvard University Press.

Wrangham, Richard W., and Dale Peterson. 1996. *Demonic Males: Apes and the Origins of Human Violence.* New York: Houghton Mifflin.

Wright, Robert. 1994. *The Moral Animal: Why We Are the Way We Are: The New Science of Evolutionary Psychology.* New York: Vintage Books.

———. 1999. *Nonzero: The Logic of Human Destiny.* New York: Pantheon

Wynne-Edwards, Vero Copner. 1962. *Animal Dispersion in Relation to Social Behaviour.* New York: Hafner Publishing Company.

Zachary, G. Pascal, and Checile Rohwedder. 2001. "Germany Widens Door for Immigrants." *Wall Street Journal,* July 2.

Zahavi, Amotz, and Avishag Zahavi. 1997. *The Handicap Principle: A Missing Piece of Darwin's Puzzle.* New York: Oxford University Press.

Zak, Paul, and Arthur Denzau. 2001. "Economics is an Evolutionary Science" In *Evolutionary Approaches in the Behavioral Sciences,* edited by Albert Somit and Stephen Peterson. Vol. 8, 31–65. Elsevier Sciences Ltd.

Zywicki, Todd J. 2000. "Was Hayek Right about Group Selection After All? Review Essay of *Unto Others: The Evolution and Psychology of Unselfish Behavior.*" *Review of Austrian Economics* 13:81–95.

Index

ABC Research Group, 158, 159–161
Abed, Riadh T., 122
abortion, 135, 147, 149, 151, 188, 189–190
Adaptively Relevant Environment, 6
Adapted Mind, The (Barkow, Cosmides, and Tooby), ix
affirmative action, 54–56, 185, 191
Afghanistan, 49–50
agrarian society, 117–118, 127, 133
agriculture, advent of, 5, 9, 102, 103
Aid to Families with Dependent Children (AFDC), 69–70
Alchian, Armen, 101
alcohol/drugs, regulation of, 145, 146, 149–150, 151, 152, 189
Alexander, Richard D., 37, 43, 61, 64, 72, 74–75, 118
allocation hierarchy, 102
alpha male, 97
Alston, Richard M., 93
altruism, xii, 143; definition of, 10; efficient, 64–67, 68, 89, 109; gene for, 23; and group selection, 57–59; kin-based, 23–24, 59; political preference for, 185–186; reciprocal, 18, 20, 59, 62, 66–67, 78, 86, 109, 162; terrorism as, 121
Ambrose, Stanley H., 6
American Law and Economics Review, xvii
American Revolution, 113, 115
analogous features, 3
Anarchy, State, Utopia (Nozick), 1
anthropomorphism, 137, 140, 151, 188
anti-Semitism, 51, 52–53, 56
apes, 3–4, 39. *See also* chimpanzees
Arnhart, Larry, 14, 130
Arrow, Kenneth J., 183, 196n
availability heuristic, 169–170
Axelrod, Robert, 60

baboons, 81
Bailey, Martin J., 75, 77

balanced polymorphism, 14
Barkow, Jerome H., 27
Barro, Robert J., 120
barter, 20
Baseñez, Miguel, 68, 84, 90, 144
base rate fallacy, 158
Baumgartner, Frank R., 163, 166
Becker, Gary, 51–52, 195n
behavior, and preferences, 14–17
Belgium, 54
Bentham, Jeremy, *see* utilitarianism
Betzig, Laura L., 102
Bingham, Paul M., 24, 41, 43, 61
Binmore, Ken, 1, 73
biological difference, gendered, 24–26
Bird-David, Nurit, 20
Boehm, Christopher, 70, 75, 78, 79, 99–100, 101, 104, 105, 106, 109, 111, 116, 118, 124, 125, 194n
bonobos, 193n
bourgeois strategy, 12, 60, 80, 86
Bowles, Samuel, 6
Boyd, Robert, 16, 42, 59, 62, 81, 102, 124, 125, 138, 140
Boyer, Pascal, 13, 83, 87, 136, 137–138, 140, 143, 144
brain size, 39, 64, 193n, 194n
bride price, 121
Brown, Donald E., 75, 77, 79
Brown, Rupert, 32
Browne, Kingsley, 26, 149
Brown v. Board of Education, 46, 55
Buchanan, James, 1, 70, 131
Burkert, Walter, 139, 148
Burundi, 45

Calculus of Consent (Buchanan and Tullock), 1
Calfee, John E., 179
Camerer, Colin, 176
Campbell, D. T., 144
capitalism, 62, 94, 125, 126, 127–128, 132–133, 175–176

free riding, 40, 67, 68, 70, 81, 88, 90
frequency dependent selection, 13, 68

game theory, 1, 11–13, 18, 132. *See also* prisoner's dilemma
Geary, David C., 97, 98, 106
gendered differences: biological, 24–26; in coalition formation, 114; in division of labor, 8, 19, 30, 116; in hierarchy, 97, 98; in mate selection, 40–41; in political behavior, 10, 113–115, 153–155; in reproductive strategy, 10; in risk behavior, 25–26, 152–155, 180
gender imbalance, 123
general-purpose intelligence, 41
genocide, 3, 38, 44, 128
Germany, 49, 54
Ghiglieri, Michael P., 3, 36
Gigerenzer, Gerd, 156–157, 158, 159–161, 196n
Gintis, Herbert, 6, 146
Glassner, Barry, 170
globalization, 110, 190
Goetze, David, 47
Goodall, Jane, 8
Good Natured: The Origins of Right and Wrong in Humans and Other Animals (De Waal), ix
gorillas, 4
government, in Western society: increasing power of, 131–132, 187; limiting power of, 127–128, 129–130, 133–134, 187, 191; need for use of force by, 129–130; role of, 129
government hierarchy, 104–105
Grady, Mark F., 82
Greif, Avner, 52
group: commonalties in, 32; definition of, 33; increase in size of, 7, 19, 27, 37–38, 153, 182
group behavior, 80–83
group good, 86
group identity, 31–34, 35–36, 37, 54, 56, 124, 135
group markers, 33
group membership mechanism, 34–35, 36, 37
group selection, 63–64, 65

Gulf War, 149
Guthrie, Stewart, 136, 137, 140

Hamilton, William, 23, 24
handicap competition, 146, 152, 189
Hanson, Jon D., 177–178
Harris, Julie A., ix–x
hawk-dove game, 11–13
Hayek, Friedrich A. von, 129–130
hierarchy, 96; allocation, 102; among chimpanzees, 97; confusion between types of, 105–108, 132; consumption, 96–100; definition of, 194n; dominance, 125–126; gendered, 97, 98; government, 104–105; in group identity, 33–34; among hunter-gatherers, 98–99, 100, 111, 116, 133; among nonhuman primates, 97; production, 100–104; reverse dominance, 78, 79, 100, 129, 187; among rhesus monkeys, 97
Hinde, Robert A., 144
hindsight bias, 158, 159, 178–179
Hinduism, 138
Hirshleifer, Jack, 49, 54, 61
Hobbes, Thomas, 1
Hoffman, Elizabeth, 84, 85
homicide, 3, 25, 99, 116, 122
Homo erectus, 5, 6, 42, 43
homologous features, 3
Homo sapiens, 3, 6
Hong Kong, 49
How the Mind Works (Pinker), ix
Hrdy, Sarah Blaffer, 131, 147
human evolution, constant elements of, 5–6
Human Nature (journal), xvii, 27
Hungary, 83
hunter-gatherers, 5; common features of, 6; conflict among, 44; division of labor among, 116; egalitarianism among, 70, 101–102, 118; food-sharing among, 94–95; group type of, 36; hierarchy among, 98–99; homicide among, 99, 116; mobile, 116–117; political power among, 116–117; polygyny among, 7; property rights among, 75, 77, 80; reverse dominance hierarchy among, 100,

About the Author

Paul H. Rubin is professor of economics and law at Emory University and editor-in-chief of *Managerial and Decision Economics.* He is a fellow of the Public Choice Society, a senior fellow at the Progress and Freedom Foundation, an adjunct scholar at the American Enterprise Institute, and former vice president of the Southern Economics Association. He was a senior economist at the Federal Trade Commission, Chief Economist at the U.S. Consumer Product Safety Commission, senior staff economist of President Reagan's Council of Economic Advisers, and has been vice president of Glassman-Oliver Economic Consultants. He has taught at the University of Georgia, City University of New York, Virginia Polytechnic Institute, and George Washington University Law School. Rubin has published more than one hundred articles and chapters in economics journals, law reviews, journals of human evolution, scholarly books, and leading newspapers. His most recent books are *Managing Business Transactions,* 1990; *Tort Reform by Contract,* 1993; *Promises, Promises: Contracts in Russia and Other Post-Communist Economies,* 1998; and *Privacy and the Commercial Use of Personal Information,* 2001 (with Thomas Lenard). He received his B.A. from the University of Cincinnati and his Ph.D. from Purdue University.